CAMBRIDGE LIBRARY COLLECTION

Books of enduring scholarly value

Technology

The focus of this series is engineering, broadly construed. It covers technological innovation from a range of periods and cultures, but centres on the technological achievements of the industrial era in the West, particularly in the nineteenth century, as understood by their contemporaries. Infrastructure is one major focus, covering the building of railways and canals, bridges and tunnels, land drainage, the laying of submarine cables, and the construction of docks and lighthouses. Other key topics include developments in industrial and manufacturing fields such as mining technology, the production of iron and steel, the use of steam power, and chemical processes such as photography and textile dyes.

Reports of the Late John Smeaton

Celebrated for his construction of the Eddystone Lighthouse near Plymouth, John Smeaton (1724–92) established himself as Britain's foremost civil engineer in the eighteenth century. A founder member of the Society of Civil Engineers, he was instrumental in promoting the growth of the profession. The Royal Society awarded him its Copley medal for research into wind and water power in 1759. After his death his papers were acquired by the president of the Royal Society, Sir Joseph Banks, Smeaton's friend and patron. Using these materials, a special committee decided to publish 'every paper of any consequence' written by Smeaton, as a 'fund of practical instruction' for current and future engineers. These were published in four illustrated volumes between 1812 and 1814. Published in 1814 using the original engraved plates, Volume 4 contains the papers that Smeaton published in the Royal Society's *Philosophical Transactions*, as well as related correspondence.

Cambridge University Press has long been a pioneer in the reissuing of out-of-print titles from its own backlist, producing digital reprints of books that are still sought after by scholars and students but could not be reprinted economically using traditional technology. The Cambridge Library Collection extends this activity to a wider range of books which are still of importance to researchers and professionals, either for the source material they contain, or as landmarks in the history of their academic discipline.

Drawing from the world-renowned collections in the Cambridge University Library and other partner libraries, and guided by the advice of experts in each subject area, Cambridge University Press is using state-of-the-art scanning machines in its own Printing House to capture the content of each book selected for inclusion. The files are processed to give a consistently clear, crisp image, and the books finished to the high quality standard for which the Press is recognised around the world. The latest print-on-demand technology ensures that the books will remain available indefinitely, and that orders for single or multiple copies can quickly be supplied.

The Cambridge Library Collection brings back to life books of enduring scholarly value (including out-of-copyright works originally issued by other publishers) across a wide range of disciplines in the humanities and social sciences and in science and technology.

Reports of the Late
John Smeaton

Made on Various Occasions,
in the Course of his Employment as a Civil Engineer

VOLUME 4: MISCELLANEOUS PAPERS,
COMPRISING HIS COMMUNICATIONS
TO THE ROYAL SOCIETY,
PRINTED IN THE *PHILOSOPHICAL TRANSACTIONS*

JOHN SMEATON

CAMBRIDGE
UNIVERSITY PRESS

University Printing House, Cambridge, CB2 8BS, United Kingdom

Cambridge University Press is part of the University of Cambridge.

It furthers the University's mission by disseminating knowledge in the pursuit of
education, learning and research at the highest international levels of excellence.

www.cambridge.org
Information on this title: www.cambridge.org/9781108069809

© in this compilation Cambridge University Press 2014

This edition first published 1814
This digitally printed version 2014

ISBN 978-1-108-06980-9 Paperback

THE

MISCELLANEOUS PAPERS

OF

JOHN SMEATON,

CIVIL ENGINEER, &c. F. R. S.

S. Brooke, Printer,
35, *Paternoster-Row, London.*

THE

MISCELLANEOUS PAPERS

OF

JOHN SMEATON,

CIVIL ENGINEER, &c. F. R. S.

COMPRISING

HIS COMMUNICATIONS TO THE ROYAL SOCIETY,

PRINTED IN THE

PHILOSOPHICAL TRANSACTIONS,

FORMING A FOURTH VOLUME TO HIS REPORTS.

ILLUSTRATED WITH PLATES.

London:

PRINTED FOR

LONGMAN, HURST, REES, ORME, AND BROWN, PATERNOSTER-ROW.

1814.

ADVERTISEMENT.

————

THE papers which are here collected together include the whole of Mr. SMEATON's communications to the Royal Society, publifhed in the Philofophical Tranfactions. Moft of the articles have been reprinted verbatim from the originals in that work : a few of them have been printed from copies which had belonged to the author, and contained corrections and additions in his own hand writing. For the latter the editor has to make his acknowledgments to Mr. SMEATON's daughter, Mrs. DIXON, who obligingly lent him the tracts.

The accompanying plates, with one or two exceptions, are the fame as were originally given with the feveral articles in the Philofophical Tranfactions ; the Prefident and Council of the Royal Society having, in the handfomeft manner, granted the loan of the coppers, in order to afford every facility in their power to the publication.

The prefent volume completes the feries of Mr. SMEATON's works. The account of the Edyftone light-houfe was publifhed during his life-time, and has been lately reprinted. Since his death, the Committee of Civil Engineers have publifhed, with the liberal permiffion of Sir JOSEPH BANKS, who had purchafed the whole of the manufcripts, three volumes of Mr. SMEATON's Reports, which comprife every paper

of

of any confequence compofed by him, in the way of his profeffion, during the long period of his public life. To thefe volumes, the prefent may be confidered as an important and indifpenfable appendage; and the whole together will be found to contain a moft valuable fund of practical inftruction in the feveral branches of Science and the Arts, which had occupied the attention of the ingenious author, and upon which he has raifed for himfelf a high and lafting reputation.

London, July 20th, 1814.

CONTENTS.

CONTENTS.

Defcription

TREATISES

AND

MISCELLANEOUS PAPERS.

[*Philos. Transact.* Vol. XLVII. Article 69.]

A LETTER from Mr. JOHN SMEATON to Mr. JOHN ELLICOTT, F. R. S. concerning some Improvements made by himself in the Air-Pump.

SIR,

Read April 16, 1752.

I HAVE been informed by fome of my friends, that my endeavours towards completing the air-pump, have been mentioned with approbation, in papers that Mr. Short and Mr. Watfon have lately communicated to the Royal Society. I underftand likewife, that the latter of thofe gentlemen has, in a very obliging manner, expreffed a wifh, that I fhould lay before the Society a particular account of my improvements therein.

I fhall always efteem it a fingular honour to be thought capable of producing any thing worthy the attention of the Royal Society; and to be my duty and intereft fo to do, upon the leaft intimation of that kind.

Your fuperior fkill in mechanics, together with the affiftance you have given me in making trial of my pump, againft three very good ones of the common conftruction, as well as the frequent marks of friendfhip you have fhewn me on all occafions, encourage me to trouble you with communicating the following to that Society, of which you are a member, and who, of all others, are the moft proper judges.

B

I fhall

I fhall not take up time with a particular recital of the alterations I have made, for near four years paft, in order to remove fome obftacles, which I imagined hindered the effects that the theory I fet out upon feemed to promife. It will be fufficient, that I give an account of what has appeared to anfwer beft, after a great number of different trials; which though fhort of what I at firft expected, yet as this pump performs much better than the common ones, my labour may not be thought wholly ufelefs; and the refpect which I have to the Society, would ftill have prevented me from troubling you or them about it at this time, could I have thought of any alteration that promifed materially to improve it.

The principal caufes of imperfection in the common pumps arife, firft, from the difficulty in opening the valves at the bottom of the barrels; and, fecondly, from the pifton's not fitting exactly, when put clofe down to the bottom; which leaves a lodgement for air, that is not got out of the barrel, and proves of bad effect, as I fhall fhew in the courfe of this paper

In regard to the firft of thefe caufes; the valves of air-pumps are commonly made of a bit of thin bladder, ftretched over a hole generally much lefs than one tenth of an inch diameter; and to prevent the air from repaffing between the bladder and the plate, upon which it is fpread, the valve muft always be kept moift with oil or water.

It is well known, that at each ftroke of the pump the air is more and more rarefied, in a certain progreffion, which would be fuch, that an equal proportion of the remainder would be taken away, was it not affected by the impediments I have mentioned: fo that, when the fpring of the air in the receiver becomes fo weak, as not to be able to overcome the cohefion of the bladder to the plate, occafioned by the fluid between them, the weight of the bladder, and the refiftance that it makes by being ftretched; the rarefaction cannot be carried farther, though the pump fhould ftill continue to be worked.

It is evident, that the larger the * hole is, over which the bladder is laid, a proportionably greater force is exerted upon it by the included air, in order to lift it up; but the aperture of the hole cannot be made very large in the common conftruction,

---

* If we examine the force, that air rarefied 140 times can exert in a common valve through a hole of one tenth of an inch diameter, we fhall find it not to exceed fix grains at a medium.

becaufe

because the pressure of the incumbent air would either burst the valve, or so far force it down into the cavity, as to prevent its lying flat and close upon the plate, which is absolutely necessary.

To avoid these inconveniences as much as possible, instead of one hole, I have made use of seven, all of equal size and shape; one being in the centre, and the other six round it; so that the valve is supported at proper distances, by a kind of grating, made by the solid parts between these holes: and to render the points of contact, between the bladder and grating, as few as possible, the holes are made hexagonal, and the partitions filed almost to an edge. As the whole pressure of the atmosphere can never be exerted upon this valve, in the construction made use of in this pump; and as the bladder is fastened in four places instead of two, I have made the breadth of the hexagons three tenths of an inch; so that the surface of each of them is more than nine times greater than common. But as the circumference of each hole is more than three times greater than common, and as the force, that holds down the valve, arising from cohesion, is, in the first moment of the air's exerting its force, proportionable to the circumference of the hole; the valve over any of these holes will be raised with three times more ease than common. But as the raising of the valve over the centre-hole is assisted on all sides by those placed round it; and as they all together contribute as much to raise the bladder over the centre-hole, as the air immediately acting under it; upon this account the valve will be raised with, at least, double the ease, that we have before supposed, or with less than a sixth part of the force commonly necessary.

It is not material to consider the force of the cohesion, after the first instant: for, after the bladder begins to rise, it exposes a greater surface to the air underneath, which makes it move more easily. I have not brought into this account the force, that keeps down the valve, that arises from the weight of the bladder, and the resistance from its being stretched; for I look upon these as small, in comparison of the other.

I was not however contented with this construction of the valves, till I had tried what effect would be produced, when they were opened by the motion of the winch, independent of the spring of the air: and though the contrivance I made use of seemed to me less liable to the objection than any thing I was acquainted with, that had been designed for that purpose; yet I did not find it to answer the end better than what I have already described; and therefore laid it aside, as it rendered the machinery much more complex, and troublesome to execute.

But

But fuppofing all thofe difficulties to be abfolutely overcome, the other defect that I mentioned in the common conftruction, would hinder the rarefaction from being carried on beyond a certain degree. For, as the pifton cannot be made to fit fo clofe to the bottom of the barrel, as totally to exclude all the air; as the pifton rifes, this air will expand itfelf; but ftill preffing upon the valve, according to its denfity, hinders the air within the receiver from coming out: hence, were this vacancy to equal the 150th part of the capacity of the whole barrel, no air could ever pafs out of the receiver, when expanded 150 times, though the pifton were conftantly drawn to the top; becaufe the air in the barrel, when in its moft expanded ftate, would be *in æquilibrio* with that in the receiver. This obftacle I have endeavoured to overcome, by fhutting up the top of the barrel with a plate, having in the middle a collar of leathers, through which the cylindrical rod works, that carries the pifton. By this means, the external air is prevented from preffing upon the pifton; but that the air, that paffes through the valve of the pifton from below, may be difcharged out of the barrel, there is a valve applied to the plate at the top, that opens upwards. The confequence of this conftruction is, that when the pifton is put down to the bottom of the cylinder, the air in the lodgement under the pifton will evacuate itfelf fo much the more, as the valve of the pifton opens more eafily, when preffed by the rarefied air above it, than when preffed by the whole weight of the atmofphere. Hence, as the pifton may be made to fit as nearly to the top of the cylinder, as it can to the bottom, the air may be rarefied as much above the pifton as it could before have been in the receiver. It follows therefore, that the air may now be rarefied in the receiver, in duplicate proportion of what it could be upon the common principle; fuppofing other impediments out of the queftion.

Another advantage of this conftruction is, that though the pump is compofed of a fingle barrel *, yet the preffure of the outward air being taken off by the upper plate, the pifton is worked with more eafe † than the common pumps with two barrels: and not only fo, but when a confiderable degree of rarefaction is defired, it will do it

---

* It is obvious that these improvements will equally obtain, whether the pump be constructed with a single or a double barrel.

† Because, though the pressure of a column of air, equal to the diameter of the piston-rod, still presses upon it, yet, as there is only the friction of one piston, and that not loaded with the weight of the atmosphere, the friction of the leather against the side of the barrel, and that of the rack and wheel, is much less: so that, notwithstanding the addition of friction in the collar of leathers, that of the whole will be less.

quicker;

quicker; for the terms of the feries expreffing the quantity of air taken away at each ftroke do not diminifh fo faft, as the feries anfwering to the common one.

I have found the gages that have been hitherto made ufe of, for meafuring the expanfion of the air, very unfit to determine in an experiment of fo much nicety. I have therefore contrived one of a different fort, which meafures the expanfion with certainty, to much lefs than 1000th part of the whole. It confifts of a bulb of glafs, fomething in the fhape of a pear, and fufficient to hold about half a pound of quick-filver. It is opened at one end, and the other is terminated by a tube hermetically clofed at top. By the help of a nice pair of fcales, I found what proportion of weight a column of mercury, of a certain length, contained in the tube, bore to that, which filled the whole veffel. By thefe means I was enabled to mark divifions upon the tube, anfwering to a 1000th part of the whole capacity, which being of about one tenth of an inch each, may, by eftimation, be eafily fubdivided into fmaller parts. This gage, during the exhaufting of the receiver, is fufpended therein by a flip-wire. When the pump is worked as much as fhall be thought neceffary, the gage is pufhed down, till the open end is immerged in a ciftern of quickfilver placed underneath : the air being then let in gradually, the quickfilver will be driven into the gage * ; till the air remaining in it becomes of the fame denfity with the external ; and as the air always takes the higheft place, the tube being uppermoft, the expanfion will be de-termined by the number of divifions, or parts of a divifion, occupied by the air at the top.

The degree, to which I have been able to rarefy the air in experiment, has generally been about 1000 times, when the pump is put clean together : but the moifture, that adheres to the infide of the barrel, as well as other internal parts, upon letting in the air, is in fucceeding trials worked together with the oil, which foon renders it fo clammy, as, being minutely divided, to obftruct the action of the pump upon a fluid fo fubtil as the air is, when fo much expanded ; but in this cafe it feldom fails to act upon the air in the receiver, till it is expanded 500 times : and this I have found it to do, after being frequently ufed for feveral months, without cleaning. I have alfo generally found it to perform beft, the firft trial at each time of ufing ; though nothing had been done at it from the time preceding ; which, after a great many trials made with

---

* The bulb of the gage may be emptied of its quicksilver, without emptying the tube ; and afterwards the tube being held horizontally, the column of mercury in it will have no power to contract or expand the air at the top.

this

this view, I alfo attribute to the watry vapours of the air mixing with the oil. An experiment, where the air was expanded 1000 times, was tried about two years fince in your prefence; at which were prefent alfo Dr. Knight and Mr. Canton; and I lately did the fame thing with Mr. Watfon *. The pump, which I have the honour of fhewing the Society, is the fame that I juft now mentioned, and the fecond that I made, with a view to improve upon this principle.

The degree of rarefaction, produced by the beft of the three pumps, that you procured the trial of, and which you efteemed good in their kind, and in complete order, never exceeded 140 times, when tried by the gage above defcribed †.

I have alfo endeavoured to render the pneumatic apparatus more fimple and commodious, by making this air-pump act as a condenfing engine at pleafure, by fingly turning a cock. This not only enables us to try any experiments under different circumftances of preffure, without changing the apparatus, but renders the pump an univerfal engine, for fhewing any effect, that arifes from an alteration in the denfity or fpring of the air. Thus, with a little addition of apparatus, it fhews the experiments of the air-fountain, wind-gun, &c.

The condenfing part is contrived in the following manner: The air above the pifton being forcibly driven out of the barrel at each ftroke, and having no where to efcape, but by the valve at the top; if this valve be connected with the receiver, by means of a pipe, and at the fame time the valve at the bottom, inftead of communicating with the receiver, be made to communicate with the external air, the pump will then perform as a condenfer.

The mechanifm is thus ordered. There is a cock with three pipes placed round it, at equal angles. The key is fo pierced, that any two may be made to commu-

---

* Though an expanfion of 1000 times is mentioned here, as capable of being procured at pleasure; yet this is by no means the utmost ever arrived at: for when perfect tightness has concurred with perfect cleanness, and perhaps some other causes, that the author does not profess to be acquainted with, it has made a rarefaction of 2000, 5000, and even 10,000 times, a degree of rarefaction seldom to be obtained in the *Torricellian* tube.

† *Boyle, Defaguliers*, and others, make mention of rarefying the air by their pumps 14,000 times and upwards: but as the experiment was made by expending a whole tube of air upon the surface of water, which is capable of emitting a quantity of air when under the exhausted receiver of the common air-pump, no certain conclusion can be drawn therefrom; on the contrary, it appears from experience, that a common air-pump that will, with certainty, rarefy the air 100 times, is not to be accounted a bad one.

nicate,

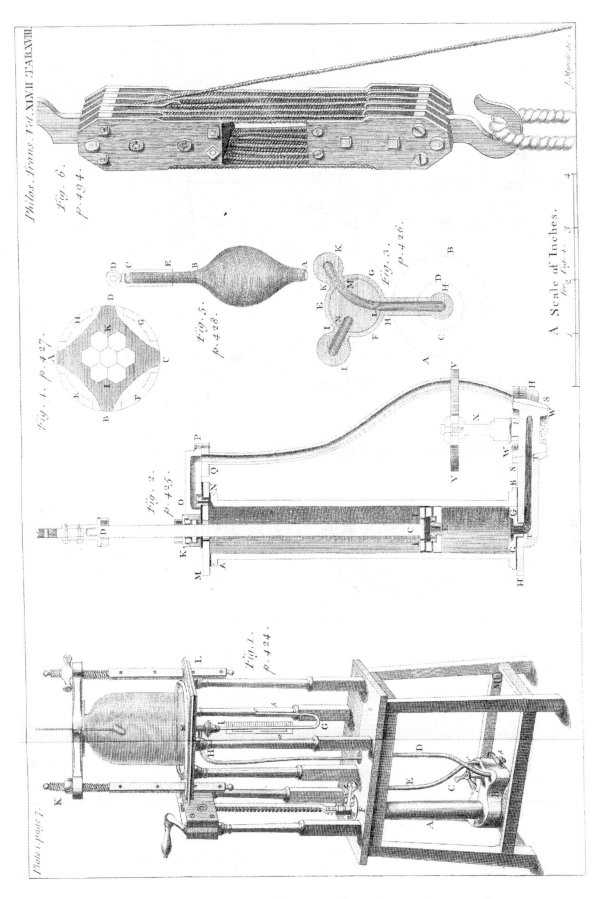

nicate, while the other is left open to the external air. One of these pipes goes to the valve at the bottom of the barrel; another communicates with the valve at the top, and a third with the receiver. Thus, when the pipe from the receiver, and that from the bottom of the barrel, are united, the pump exhausts: but turn the cock round, till the pipe from the receiver, and that from the top of the barrel, communicate, and it then condenses. The third pipe, in one case, discharges the air taken from the receiver, into the barrel; and in the other, lets it into the barrel, that it may be forced into the receiver.

I am,

SIR,

Your most humble servant,

J. SMEATON.

Furnival's-Inn Court,
April 16, 1752.

P. S. I have also added some draughts, and letters of reference, in order to explain myself more fully.

Plate 1. Figure 1.

Is a perspective view of the principal parts of the pump together.

A is the barrel.

B the cistern, in which are included the bottom of the barrel and cock, with three communications. These are covered with water to keep them air-tight. A little cock to let the water out of the cistern, is marked *b*.

C *c c* is the triangular handle of the key of the cock: which, by the marks on its arms, shews how it must be turned, that the pump may produce the effect desired.

D H is the pipe of communication between the cock and the receiver.

E is the pipe, that communicates between the cock and the valve, on the upper plate of the barrel.

F is the upper plate of the pump, which contains the collar of leathers *d*, and V, the valve, which is covered by the piece *f*.

G I

G I is the fiphon-gage; which fcrews on and off, and is adapted to common purpofes. It confifts of a glafs tube, hermetically fealed at c, and furnifhed with quickfilver in each leg; which, before the pump begins to work, lies level in the line $a b$;
the fpace $b c$ being filled with air of the common denfity. When the pump exhaufts,
the air in $b c$ expands, and the quickfilver in the oppofite leg rifes, till it becomes a
counter-balance to it. Its rife is fhewn upon the fcale I a, by which the expanfion of
the air in the receiver may be nearly judged of. When the pump condenfes, the
quickfilver rifes in the other leg, and the degree may be nearly judged of, by the contraction of the air in $b c$: marks being placed at $\frac{1}{2}$ and $\frac{1}{3}$ of the length of $b c$ from c;
will fhew when the receiver contains double or treble its common quantity.

K L is a fcrew-frame to hold down the receiver, in condenfing experiments, which
takes off at pleafure; and is fufficient to hold down a receiver, the diameter of whofe
bafe is feven inches, when charged with a treble atmofphere: in which cafe it acts
with a force of about 1200 pounds againft the fcrew-frame.

M is a fcrew that faftens a bolt, which flides up and down in that leg, by means
whereof the machine is made to ftand faft on uneven ground.

Fig. 2.

Is a perpendicular fection of the barrel and cock, &c. where

A B reprefents the barrel.

C D the rod of the pifton, which paffes through

M N the plate, which clofes the top of the barrel.

K is the collar of leathers, through which the pifton-rod paffes. When the pifton
is at the bottom of the cylinder, the upper part of K is covered by the cap at D, to
keep out duft, &c.

The valve on the upper plate, is covered by the piece

O P, which is connected with the pipe

Q H, which makes the communication between the valve and cock.

C E

C E is the pifton ; and

E F F is the pifton-valve.

I I are two little holes to let the air pafs from the pifton-valve into the upper part of the barrel.

G G T is the principal valve at the bottom of the cylinder.

H H is a piece of metal, into which the valve G G T is fcrewed, and clofes the bottom of the cylinder; out of which alfo is compofed

S S the cock, and

T T the duct from the cock to the bottom of the barrel.

W W is the key of the cock.

X the ftem; and

V V the handle.

Fig. 3.

Is an horizontal fection of the cock, through the middle of the duct T T.

A B reprefents the bignefs of the circular plate, that clofes the bottom of the barrel.

C D reprefents the bignefs of the infide of the barrel.

E F G is the body of the cock; the outward fhell being pierced with three holes, according to an equilateral triangle, and correfponding to the three ducts L H, N I, M K, whereof

L H is the duct, that goes to the bottom of the barrel.

C

N I,

N I, the duct, that communicates with the top of the barrel; and

I I the flanch or base for the joint, with the pipe E, in fig. 1.

M K is the duct, that passes from the cock to the receiver; and

K K the flanch for the joint, with the pipe D, in fig. 1.

L M N is the key, or solid part of the cock, moveable round in the shell E F G. When the canal L M answers to the ducts H and K, the pump exhausts, and the air is discharged upwards by the perforation N. But the key L M N being turned till the canal L M answers to I and K, the perforation N will then answer to H H; and in this case the pump condenses. Lastly, when N answers to K K, the air is then let in or discharged from the receiver, as the circumstance requires.

Fig. 4.

Is the plan of the principal valve.

A B C D represents the bladder fastened in four places, and stretched over the seven holes I K, formed into an hexagonal grating; which may be called the honeycomb.

E F G H shews where the metal is a little protuberant, to hinder the piston from striking against the bladder.

Fig. 5.

Represents the new gage; which, from its similitude, may be called the pear-gage. It is open at A; B C is the graduated tube, which is hermetically closed at C, and is suspended by the piece of brass D E, that is hollowed into part of a cylinder, and clasps the tube.

AN

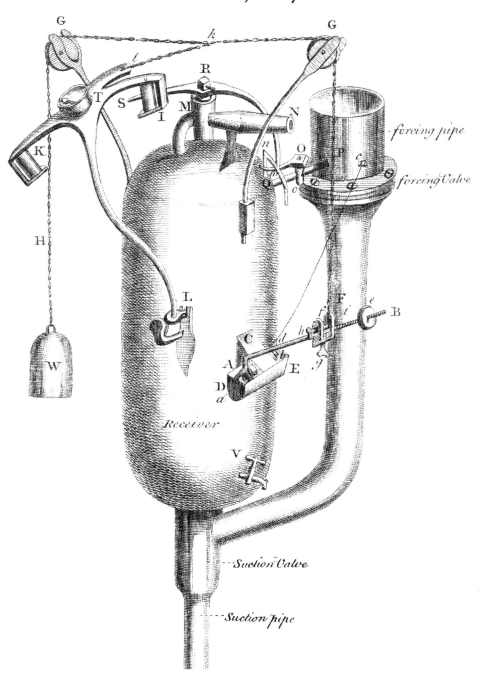

forcing pipe

forcing Valve

Receiver

Suction Valve

Suction pipe

J. Mynde sc.

AN ENGINE for raising Water by Fire, being an Improvement of SAVERY's Construction, to render it capable of working itself, invented by Mr. DE MOURA, of Portugal, F. R. S. described by Mr. J. SMEATON.

Read Nov. 9, 1752. THIS engine confists of a receiver, a fteam, and an injection cock, a fuction and a forcing pipe, each furnifhed with a valve, together with a boiler, which, on account of its bulk and weight, is not fent with the reft; but, as it may be of the common globular fhape, and having nothing particular in its conftruction, a defcription of it will not be neceffary, as alfo the reft of thefe parts already mentioned being effential to every machine of this kind, a further account of them may be difpenfed with. What is peculiar to this engine is a float within the receiver, compofed of a light ball of copper, which is not loofe therein, but faftened to the end of an arm, which is made to rife and fall by the float, while the other end of the arm is faftened to an axis, and, confequently, as the float moves up and down, the axis is turned round one way or the other. This axis is made conical, and paffes through a conical focket, which laft is foldered to the fide of the receiver. Upon one of the ends of the axis, which projects beyond the focket, is fitted a fecond arm, which is alfo moved backwards and forwards by the axis, as the float rifes or falls. By thefe means, the rifing or falling of the furface of the water within the receiver, communicates a correfpondent motion to the outfide, in order to give proper motions to the reft of the geer; which regulates the opening and fhutting of the fteam and injection cocks, and ferves the fame purpofe as the plug frame, &c. in Newcomen's engine. The particular conftruction and relation of thofe pieces will better appear by the figure and references, than can be done by a general defcription.

Plate 2.

A B an arm, which is faftened to

a b, a conical axis, which goes through a conical focket in

C, a triangular piece foldered to the receiver. This piece has this fhape, to give liberty to the arm to rife and fall; that carries the float on the infide.

D E is a fmall ciftern, foldered to the receiver, which, being kept full of water, keeps the axis and focket air-tight. This ciftern is conftantly kept full of water by

means

means of a fmall leakage through the wooden peg *c,* which follows the packthread *c d* to the ciftern.

e, is a fmall weight to counterpoife the float within.

f, is a flider; which being fet nearer to or farther from the axis, will rife or fall, a greater or leffer fpace as may be required, and is faftened by the fcrew *g.* This flider is furnifhed with a turn about, *h i,* which is alfo faftened by a fcrew and nut at the end *i,* and ferves to adjuft the length of

F G G H, a chain, which gives motion, by means of the fhorter chain *k l,* to

I K L, the balance which opens and fhuts the cocks; and moves upon the fmall axis L.

G G are two pullies, fupported by two arms, that are faftened to the fide of the receiver, and give the chain a proper direction in order to move the balance.

M N is the fteam cock, the end N being fuppofed to be detached from a pipe, that gives it communication with the boiler.

O is the injection cock, whofe key is turned by the arm O *m.*

P Q is the injection pipe, communicating between the forcing pipe above the valve, and the top of the receiver.

R S is the arm, by which the key of the fteam-cock is worked.

I K, two rollers annexed to the balance, which, by ftriking upon the arm R S, open and fhut the fteam cock, as the balance is moved backward and forward.

R *n o* is the fteam cock's key tail, which is furnifhed with two fmall rollers, *n o,* which open and fhut the injection cock, by acting upon the arm O *m,* in fuch a manner, that when the fteam cock is opened, the injection is fhut, and *vice verfâ.*

T is a bell of advice, which, moving along with the balance, continues to ring as long as the engine is at work.

V is

V is a cock, which ferves to difcharge the air from the receiver, and is opened by hand when neceffary.

W is a weight fufficient to raife the balance to a perpendicular pofture, when it is inclined to the right, and alfo to overcome the friction of the float, axis, pullies, chain, &c.

To put the engine in motion, prefs down the arm A B, which will bring the balance over to the right fide, and its motion will open the fteam cock, and fhut the injection; fet open the cock at V, that the air may be difcharged by the entrance of the fteam into the receiver. This being done, fhut that cock, and let go the arm; the weight W will bring over the balance to the left, and in its motion fhut the fteam cock, and open the injection; the prefently condenfing the fteam into water, in a great meafure leaves a vacuum in the receiver. Things remain in this fituation, till the preffure of the atmofphere has caufed the water to mount through the fuction pipe into the receiver, where, as its furface rifes, it caufes the float to afcend; and depreffing the arm A B, raifes the balance, till it has paffed the perpendicular; and, in its defcent, which is done by its own gravity, the roller K lays hold of the arm R S, again opens the fteam cock, and fhuts the injection. The receiver being now almoft filled with water, the balance cannot return till the furface of the water therein fubfides, and fuffers the float to defcend. This is performed by the elafticity of the fteam, which, at the fame time that it fills the receiver, drives out the water through the forcing pipe, and when the furface is defcended fo low, as to fuffer the weight W to bring the balance beyond the perpendicular towards the left, it then falls of its own accord, and, in falling, the roller I, lays hold of the arm R S, fhuts the fteam cock and opens the injection, as before.

When the engine is defired to be ftopped, obferve, when the balance lies to the right, to turn round the arm O m of the injection cock, fo that the tail of the fteam cock may mifs it in the next motion; fo that, at the fame time that the receiver is filled with fteam, and the fteam cock fhut, the injection not being opened, the motion will ftop for want thereof.

AN

AN Account of some Experiments upon a Machine for measuring the Way of a SHIP at Sea.

IN the *Philosophical Transactions*, No. 391, for November 1725, Mr. Henry de Saumarez, gives an account of a machine for measuring a ship's way more exactly than by the log. This machine consists of a first mover, in the form of the letter Y. Upon the two arms of the Y are fastened two vanes, inclined in such a manner, that when the Y is hauled through the water by a rope, fastened to the stem or tail thereof, it may turn round, and, of consequence, endeavour to turn the rope round. The other end of the rope being fastened to the end of a spindle capable of moving freely round, will be made to do so by the rotations of the Y, communicated to the rope. A motion being thus communicated to a spindle within the ship, this spindle may be made to drive a set of wheel-work, which will register the turns of the Y; and the value of a certain number of these turns being once found, by proper experiments, they are easily reducible into leagues and degrees, &c. The only difficulty then is, whether this Y will make the same number of rotations in going the same space, when it is carried through the water fast, as when it is carried flow. Upon this head Mr. de Saumarez, as well in the paper above cited, as in a subsequent one published in the *Philosophical Transactions*, No. 408, for March 1729, has given an account of several trials, which he has made of it, from which it appears, that this machine, in part, answers the end proposed, and is, in part, defective: the errors of which he supposes to proceed from the sinking down of the Y into the water, upon a flow motion; the axis of its rotation being then more oblique to the horizon than in a quick one.

In a machine, constructed like this, it is evident, that the end of the spindle, to which the rope is fastened, must be of sufficient strength and thickness, not only to bear the force or stress, that the hauling of the Y through the water will lay upon it, in the greatest motion of a ship, but also to bear the accidental jerks, that the waves will superadd thereto. The thickness of the spindle then being determined by these conditions; it is also manifest, that, to prevent the spindle from being pulled out of its place by the draft of the rope, there must be a shoulder formed upon it, which must be greater than the part of the spindle before described, for the spindle to bear against. The size, that Mr. Saumarez proposes to give to his Y, is twenty-seven

inches

inches the whole length; fifteen inches for the length of the arms (which are to be opened to a right angle); eight inches for the length of each vane; four inches and a half broad, and the ftems and fhank to be two-thirds of an inch thick. According to thefe dimenfions, the refiftance, that this part of the machine will meet with, in paffing through the water, will, in the fwift motions of the fhip, be very confiderable: confequently, the neceffary bulk of the pivot-end of the fpindle, and its fhoulder, will occafion a confiderable friction in the turning thereof, and retardation to the rotation of the machine.

To cure thefe defects as much as poffible, inftead of the **Y** before defcribed, I made trial of a fingle plate of brafs, of about ten inches long, two and an half broad, one-thirtieth of an inch thick, and cut into an oval fhape. This plate being fet a little a-twift, and faftened by one end to a fmall cord, in the manner of the **Y**, is likewife capable of making a rotation, in being drawn through the water; but with this difference, that as this is but a fmall thin plate drawn edgeways through the water, its refiftance in paffing through it is much lefs; of confequence, a much fmaller line is fufficient to hold it, which again confiderably diminifhes the refiftance; and this, of courfe, proves a double diminution of friction in the fpindle: Firft, as the preffure upon it is lefs; and, fecondly, as it allows the fpindle and fhoulder to be of a lefs diameter. To break the jerks of the waves; next to the end of the fpindle I fixed a fpiral fpring of wire, to which the cord was faftened; which, by this means, was capable of playing backwards and forwards, and giving way to the irregularities of the fea: and, left the plate fhould lay faft hold of any thing, or any extraordinary jerk fhould damage the fpindle or fpring, a knob, or button, was faftened upon the cord, at a fmall diftance from the fpring, which ftopped upon a hole in a piece of wood, and prevented the fpring from being pulled out to above a certain length; fo that all addition of force, beyond this, could only tend to break the cord, and carry away the plate. The fpindle, being thus guarded from accidents, will allow of a ftill further diminution of its fize; fo that, at laft, I ventured to make the fpindle-pivot no more than one-twentieth of an inch diameter, and that of the fhoulder one-eighth; being of tempered fteel, and fufficiently fmooth. The hole, in which the pivot, and againft which the fhoulder worked, was of agate likewife, well polifhed.

Being thus provided, in May 1751, I procured a boat, upon the ferpentine river in Hyde-park, to try how far the turns of the machine would be confiftent with themfelves, when the fame fpace was meafured over with the fame, and with different velocities.

velocities. The courfe was determined at each end, by obferving the co-incidence of two trees, in a line nearly at right angles to the river. We, however, rowed beyond the mark, that the machine might be in full play when the courfe was begun: the fpindle was ftopped at the beginning and end; the numbers read off, and were as follows:

The fpace between the marks was, by eftimation, about half a mile.

				Revol[s]
1ft rowing up the river, in 11 minutes, the plate made	.	.	.	615
2d down 14	. . . . .	.	.	645
3d up 18 and an half	. . . .	.	.	612
4th down 9 and an half	. . . .	.	.	603
5th up 18	. . . . .	.	.	620
6th down 10	. . . . .	.	.	600

It is obfervable, that the greateft difference among the above obfervations, is between the 2d and 6th, being 645 and 600; the difference being about one-fourth part of the whole; the times being 14 minutes and 10, both in going down the river: whereas thofe obfervations, which differ moft in point of time, viz. the 3d and 4th, being performed in 18 minutes and an half, and 9 minutes and an half, refpectively; have their revolutions more nearly alike, being 612 and 603; which differ only by one fixty-eighth of the whole. From thefe obfervations I was led to think, that the different velocities, wherewith a veffel moves forwards, would make no material difference in the number of rotations of the plate; or, at leaft, that thofe differences would be lefs than the irregularities arifing from other caufes, even in trials nearly fimilar.

The next trial of this machine was on board a fmall failing veffel, in company with Dr. Knight, and Mr. William Hutchinfon, an experienced feaman, and mafter of a confiderable merchant fhip. Our expedition was upon the river Thames, and fome leagues below the Nore. The intention of the trial here was, to find, in general, how far it agreed with the log, and how it would behave in the fwell of the fea; a comparifon with the meafure of a real diftance being here impracticable, on account of the tides and currents.

The method of trial was this: We fuffered the whole log line to run out, being 357 feet between the firft knot and the end. The perfon, who hove the log, gave notice,

at

at the extremes of this meafure, that the perfon, who attended the dial of the machine, might ftop the fpindle at the beginning and end; while a third obferved, by a feconds-watch, the time taken up in running thefe 357 feet.   By thefe means, we were enabled to afcertain the comparative velocity wherewith we moved, and the number of turns of the plate at each trial, correfponding to 357 feet by the log; which, if the machine and log were both accurate, ought to have been always the fame.   The particulars of thefe experiments are contained in the following table:

Turns of the
  Plate.

83	In the river at anchor by the tide	. .	124
82	The fame repeated	. . . .	134
81	Sailing in the river	. . . .	98
79	In the river at anchor by the tide	. .	135
76	Sailing in the river	. . . .	115
74	At fea upon a wind	. . . .	64
74	The fame repeated	. . . .	69
71	Sailing in the river	. . . .	71
70	The fame	. .	66
70	Before the wind at fea	. . .	77
70	The fame	. .	56
70	The fame	. .	52
66	Before the wind in the river	. . .	55
64	The fame	. .	53
64	The fame	. .	60
64	The fame	. .	43
63	At fea upon a wind	. . . .	53
62	The fame	. .	52
62	Sailing in the river	. . . .	45

Seconds of time during the running out of 357 feet of log-line.

It appears from thefe trials, made in different pofitions of the veffel with regard to the wind, both in the river and at fea, as well by the tides at anchor, as in failing, that the turns of the plate, correfponding with the fpace of 357 feet by the log, were from 62 to 83; and the times, in which this fpace was run, were from 45 to 135 feconds; the greater number of revolutions anfwering to the greater number of feconds, or flower movement of the veffel.   Upon finding this confiderable difagreement between the log and plate, when fwift and flow motions are compared, I did not

D   fuppofe, suppose, that they proceeded from a retardation of the plate in swift motions, but from the hauling home of the log in slow ones. As for instance; the log, to do its office accurately, ought to remain at rest in the water, whatever be the motion of the vessel. But even the keeping the line strait, and much more the suffering the log to haul the line off the reel (as practised by many,) will make the log, in some measure, follow the vessel, and will be greater, in proportion as the time of continuance of this action is greater; and therefore the log will follow the ship twice as far in going one knot, when the ship is twice as long in running it. The consequence of this is, that a vessel always runs over a greater space than is shewn by the log-line; but that this error is greater, in proportion as the vessel moves slower. It is this reason, I suppose, that has induced the practical seamen to continue the distance between their knots shorter than they are directed by the theory.

Afterwards, in the same summer, I made such another expedition, in a sailing vessel, along with Captain Campbell, of the Mary yacht, and Dr. Knight. Having prepared two of these machines as near alike as possible, I determined to try, how far they were capable of agreement, when exposed to the same inconveniencies, and used together. During the trial of these machines, one made 86,716 revolutions, and the other made 88,184. During this space, they were compared at ten several intervals. The revolutions between each interval differed from the proportion of these numbers, in the first comparison, one-nineteenth of the whole interval. The errors of each interval, in the other comparisons, were, in order, two-seventeenths, one-nineteenth, one-twentieth, one-fifty-fourth, one-fourteenth, one-eightieth, one-sixty-seventh, one-fourteenth, one-sixteenth; the greatest errors being where the spaces were the shortest. In other respects, the plates seemed to perform their duty, in the water, well enough, though the sea was as rough, in this voyage, as our small vessel would well bear.

Lastly, being for some time on board the Fortune sloop of war, commanded by Alexander Campbell, Esquire; in company with Dr. Knight, for the purpose of making trial of his new invented sea-compasses, I had frequent opportunities of making use of these machines, by comparing them with one another, with the log, and with real distances; and having, by repeated trials, pretty well ascertained the number of turns of the plate, that was equal to a given space, by the help of the log, in the manner before described, when the ship was upon a middle velocity; I found the spaces, so measured, nearly consistent with themselves, and with the truth: but all this while the winds and weather were very moderate. It afterwards happened, that we run eighteen leagues in a brisk gale of wind, which, though not fair for us (being

before

before the beam,) yet drove us fometimes at the rate of eight knots an hour, as appeared by heaving the log. During this run I obferved, that the refiftance of the water, to the line and plate, was very confiderable, and increafed the friction of the fpindle fo much, as to prevent it from beginning to turn, till the plate had twifted the line to fuch a degree, that when it did fet a going, it would frequently run 150 or 200 turns at once. I alfo obferved, that the wind coming acrofs the courfe of the fhip, blew the cord a good deal out of the direction of the fpindle, and caufed the line to rub againft the fafeguard hole, for the button to ftop againft, as above defcribed; which undoubtedly occafioned confiderable friction in that place. But the moft untoward circumftance that I obferved, was, that being in a rough, but fhort chopping fea, and failing obliquely acrofs the waves, the plate would frequently be drawn from one wave to another through the air, without touching the water; and, as it appeared, would jump from one wave to another, the unevennefs of the furface, joined to the quicknefs of the motion, not permitting the plate to follow the depreffion of the water. This evil I endeavoured to remedy, by placing upon the line, at a fmall diftance before the plate, fome hollow bullets, fuch as are made for nets, in order to keep the plate fo low down in the water, as to be below the bottom of the waves. This, in part, I found they did; but they, at the fame time, added fo much refiftance, in their paffing through the water, that the inconvenience was as great one way as the other.

Upon making up the account of this run, I found the number of rotations were lefs, by one full third than they ought to have been, compared with former obfervations; which afforded me a convincing proof, that this inftrument was confiderably retarded in quick motions.

The length of the line made ufe of was about twenty fathoms, which I found neceffary, that the water, difturbed by the body of the fhip, might be tolerably fettled before the plate was drawn through it; but this length of line was alfo an inconvenience, as it met with greater refiftance in the water.

Upon the whole, it feems to me, that an inftrument, made as above defcribed, is capable of meafuring the way of a fhip at fea, when its velocity does not exceed five fea miles an hour, to a degree of exactnefs exceeding the log. It therefore may be ufeful in the menfuration of the velocities of tides, currents, &c. and alfo in meafuring diftances at fea in taking furveys of coafts, harbours, &c. Thus far it feems capable of performing, upon the fuppofition, that it cannot be brought to a greater degree of perfection. But this I am very far from fuppofing; on the contrary, I do

not

not defpair, that it may be brought to anfwer the end of meafuring the way of a fhip at fea univerfally; and, for that reafon, it may not be amifs to put down a few hints, concerning the caufe and cure of the errors above-mentioned, for the fake of thofe, who may hereafter be inclined to profecute thefe inquiries *.

It appears then from the preceding obfervations, that the rotation of the plate is confiderably retarded in the quickeft motions of the fhip; and fenfibly fo, in all velocities exceeding five miles an hour. This may proceed, firft, from the friction of the machine increafing in a greater proportion than the power to turn it round. Secondly, from the water's being put in motion by the fhip, fo as to follow it in the fame direction, and that to a confiderable diftance aftern. And, thirdly, from the plate's jumping from wave to wave, when their concavity is great and diftance little.

The firft may, in fome meafure, be helped, by applying a loaded fly, of a proper fize, to the fpindle of the machine, which will prevent its fticking faft for a time, and then whirling round with great rapidity, as it is apt to do when the refiftance is great; by which means, the motion will be rendered more equal and uniform, as was juftly obferved to me by my friend Mr. Ellicott, of this Society.

Alfo, if the body of the machine were hung, equally poifed upon crofs-centres, like thofe ufed for fea-compaffes, or in the manner of a fwivel-gun, as Captain Alexander Campbell well propofed; the fpindle of the machine would readily place itfelf in the fame direction with the line that draws it, and thereby avoid unneceffary frictions from the oblique direction of the cord.

The fecond may be helped by placing the machine upon the end of a pole, faftened near the forecaftle, over the fide of the fhip. By this means, a fhorter line will be neceffary, and the plate prevented from working in the more difturbed water at the ftern.

Laftly, its quitting the water, perhaps, might be helped by joining a fhank of brafs, of fix inches long, and three quarters of an inch diameter, to the fore-part of

---

* Upon communicating thefe experiments and obfervations to my ingenious friend Mr. William Ruffel, he gave me an account of a machine, that he had made trial of in a voyage, fome years fince, from the Levant, fo nearly agreeing with the above defcribed, that one would have imagined we had been of each other's council in defigning them.

the

the plate, to which the cord muſt be faſtened, the ends of the ſhank being formed into a figure moſt convenient for paſſing through the water with eaſe. The weight of this will cauſe the fore-part of the plate to ſink faſter than the other, and endeavour to give it a direction down into the water *.

I had intended to have made trial of the effect of theſe alterations, but have been prevented, partly by want of opportunity, and partly from the indifference, with which I found ſuch a contrivance as this, even if brought to perfection, was likely to be received by ſeamen; who, in general, do not ſeem to be over fond of making trial of new inſtruments, eſpecially if propoſed by landmen, as, in deriſion, they are pleaſed to call us.

Indeed it may be objected, that, could we meaſure the way of a ſhip through the water ever ſo exactly, unleſs ſome method were found out, of aſcertaining the currents, &c.; a ſhip's courſe, with reſpect to the globe, could not hereby be determined. But then it may be replied, with equal juſtice, that it is for want of a means of meaſuring the way of a ſhip through the water (and this compared with other check obſervations,) that the drift and velocities of the principal currents have not already been determined.

Mr. de Saumarez, in his ſecond paper, of March 1729, makes mention of another machine for this purpoſe, which he himſelf acknowledges to be inferior to his former, eſpecially in rough weather at ſea. But as ſeveral others have fallen upon, and propoſed, a machine ſimilar to this; it may not be amiſs to add the following remarks upon it. The firſt mover, in this, is compoſed of four arms, fixed to the bottom of a perpendicular ſpindle; each arm is furniſhed with a vane, which opens one way, and ſhuts the other, as ſome have attempted the making of horizontal wind-mills. This, by being carried through the water progreſſively, will turn round, and the faſter, as the ſhip moves faſter: but to judge, whether it will do it proportionably in all velocities of the ſhip, let us conſider,

1. That a good ſailing ſhip will frequently ſail at the rate of ten ſea miles (ſixty to a degree) an hour, which is at the rate of ſeventeen feet *per* ſecond.

---

* Mr. Ruſſel's plate was joined to a shank, who never found it to jump out of the water, at any time, when he made uſe of it.

2. Suppoſing

2. Suppoſing the ſide of the fly, where the vanes are cloſed, to be retained by the water at reſt; the oppoſite ſide of the fly, where the vane is open, muſt meet the water with a velocity double to that of the ſhip, or at the rate of thirty-four feet in a ſecond: as would be the caſe with the upper part of a coach-wheel, whoſe velocity through the air is double to that, wherewith the coach moves forward.

3. That a plane ſurface of three inches ſquare, moving through the water with a velocity of thirty-four feet *per* ſecond, will meet with a reſiſtance, at leaſt, equal to ſeventy pounds avoirdupoiſe.

4. That the reſiſtance, which the open vanes will meet with in the water, will, in ſwift motions, be very conſiderable, and, of conſequence, the fly will move much ſlower than it ought to do, if theſe reſiſtances were leſs.

5. That from hence there is much reaſon to doubt, whether the reſiſtance of the medium, and friction of the machine, taken together, will always produce ſuch diminu-tion, in the number of turns, as that the number of revolutions, actually ſhewn by the indexes, may be the ſame when the ſame ſpace is gone over with a great velocity, as with a ſmall one.

AN

AN ACCOUNT of some Improvements of the Mariner's Compass, in order to render the Card and Needle proposed by Dr. KNIGHT, of general Use, by JOHN SMEATON, Philosophical Instrument Maker.

Prefented July 5th, 1750, and printed, with Alterations.

THE cover of the wooden box being taken off, the compafs is in a condition to be made ufe of in the binacle, when the weather is moderate: but if the fea runs high, as the inner box is hung very free on its centres, (the better to anfwer its other purpofes,) it will be neceffary to flacken the milled nut, placed upon one of the axes that fupports the ring, and to tighten the nut on the infide that correfponds to it. By this means the inner box and ring will be lifted up from the edges, upon which they reft when free; and the friction will be increafed, and that to any degree neceffary to prevent the too great vibration, which otherwife would be occafioned by the motion of the fhip.

To make the compafs ufeful in taking the magnetic azimuth, or amplitude of the fun and ftars, as alfo the bearing of headlands, fhips, and other objects, at a diftance, the brafs edges, defigned at firft to fupport the card, and throw the weight thereof as near the circumference as poffible, is itfelf divided into degrees and halves; which may be eafily eftimated into fmaller parts, if neceffary.

The divifions are determined by means of a cat-gut line, ftretched perpendicularly with the box, as near the brafs edge as may be, that the parallax arifing from a different pofition of the obferver may be as little as poffible.

Underneath the card are two fmall weights, fliding on two wires, placed at right angles to each other, which being moved nearer to, or farther from the centre, counterbalance the dipping of the card in different latitudes, or reftore the *equilibrium* of it, where it happens by any other means to be got too much out of level.

There is alfo added an *index* at the top of the inner box, which may be put on and taken off at pleafure, and ferves for all altitudes of the object. It confifts of a bar

equal

equal in length to the diameter of the inner box, each end being furnifhed with a perpendicular ftile, with a flit parallel to the fides thereof. One of the flits is narrow, to which the eye is applied, and the other is wider, with a fmall cat-gut ftretched up the middle of it, and from thence continued horizontally from the top of one ftile to the top of the other: there is alfo a line drawn along the upper furface of the box. Thefe four, viz. the narrow flit, the horizontal cat-gut thread, the perpendicular one, and the line on the bar, are in the fame plane, which difpofes itfelf perpendicular to the horizon, when the inner box is at reft, and hangs free. This index does not move round, but is always placed on fo as to anfwer the fame fide of the box.

When the fun's azimuth is defired, and his rays are ftrong enough to caft a fhadow, turn about the wooden box, till the fhadow of the horizontal thread, or (if the fun be too low) till that of the perpendicular thread in one ftile, or the light through the flit in the other, falls upon the line on the index bar, or vibrates to an equal diftance on each fide of it, gently touching the box, if it vibrates too far: obferve at the fame time the degree marked upon the brafs edge by the cat-gut line. In counting the degree for the azimuth, or any other angle that is reckoned from the meridian, make ufe of the outward circle of figures upon the brafs edge, and the fituation of the index bar, with regard to the card and needle, will always direct upon what quarter of the compafs the object is placed.

But if the fun does not throw out fufficiently ftrong, place the eye behind the narrow flit in one of the ftiles, and turn the wooden box about, till fome part of the horizontal or perpendicular thread appears to interfect the centre of the fun, or vibrate to an equal diftance on each fide of it, ufing fmoked glafs next the eye, if the fun's light is too ftrong. In this method another obferver will be generally neceffary to note the degree out by the *nonius*, at the fame time the firft gives notice that the thread appears to fplit the object.

From what has been faid, the other obfervations will be eafily performed; only in cafe of the fun's amplitude, take care to number the degrees by the half of the inner circle of figures on the card, which are the complements of the outer to ninety, and confequently fhew the diftance from eaft or weft.

The azimuth of the ftars may alfo be obferved by night: a proper light ferving equally for one obferver to fee the thread, and the other the degree upon the card.

It

Plate 3 Page 25.

MARINERS COMPASS.

Fig.1.

Fig.4.

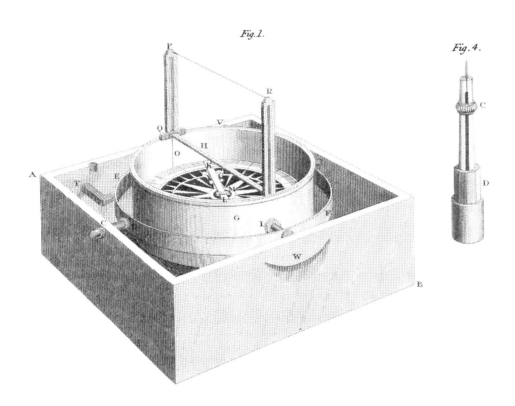

Fig. 2.

Fig.3.

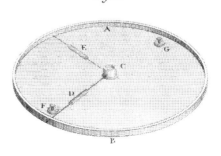

It may not be amiſs to remark farther, that, in caſe the inner box ſhould loſe its *equilibrium*, and conſequently the index be out of the plane of a vertical circle, an accurate obſervation may ſtill be made, provided the ſun's ſhadow is diſtinct: for, by obſerving firſt with one end of the index towards the ſun, and then the other, a mean of the two obſervations will be the truth.

Explanation of the figures, PLATE III.

Fig. 1, is a perſpective view of the compaſs, when in order for obſervation. The point of view being the centre of the card, and the diſtance of the eye two feet.

A B is the wooden box.

C and D are two milled nuts; by means whereof the axis of the inner box and ring are taken from the edges on which they move, and the friction increaſed when neceſſary.

E F is the ring that ſupports the inner box.

G H is the inner box; and

I is one of its axis, by which it is ſuſpended on the ring E F.

K L is the magnet, or needle; and

M a ſmall brace of ivory, that confines the cap to its place. See fig. 2.

The card is a ſingle varniſhed paper, reaching as far as the outer circle of figures, which is a circle of thin braſs, the edge whereof is turned down at right angles to the plane of the card, to make it more ſtiff.

O is a catgut line drawn down the inſide of the box; for determining the degree upon the braſs edge.

P Q R S is the index bar, with its two ſtiles and catgut threads; which, being taken off from the top of the box, is placed in two pieces T and V, notched properly to receive it.

W is a piece cut out in the wood, ſerving as a handle.

E

Fig. 2.

Fig. 2, is the card *in plano*, with the needle fixed upon it; being one third of the diameter of the reel card.

Fig. 3, is a perfpective view of the back fide of the card, where

A B reprefents the turning down of the brafs edge.

C is the under part of the ivory cap.

D and E are the two fliding weights to balance the card; and

F and G, two fcrews that fix the brafs edge, &c. to the needle.

Fig. 4, is the pedeftal that fupports the card, containing a fewing needle, fixed in two fmall grooves to receive it, by means of the collet C, in the manner of a port crayon.

A D, the ftem, is filed into an octagon, that it may be the more eafily unfcrewed.

AN

Plate 4. page 27.

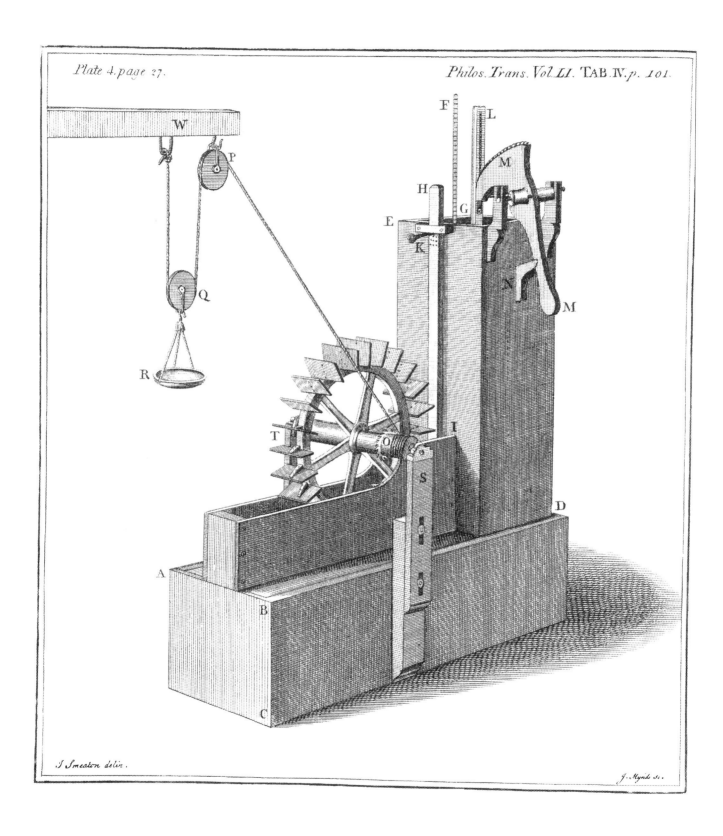

AN Experimental Enquiry concerning the Natural Powers of Water and Wind to turn Mills, and other Machines, depending on a Circular Motion.

Read before the Royal Society, May 3d and 10th, 1759.

WHAT I have to communicate on this fubject was originally deduced from experiments made on working models, which I look upon as the beft means of obtaining the outlines in mechanical enquiries. But in this cafe it is very neceffary to diftinguifh the circumftances in which a model differs from a machine in large; otherwife a model is more apt to lead us from the truth than towards it. Hence the common obfervation, that a thing may do very well in a model that will not anfwer in large. And, indeed, though the utmoft circumfpection be ufed in this way, the beft ftructure of machines cannot be fully afcertained, but by making trials with them, when made of their proper fize. It is for this reafon that though the models referred to, and the greateft part of the following experiments, were made in the years 1752 and 1753, yet I deferred offering them to the Society, until I had an opportunity of putting the deductions made therefrom in real practice, in a variety of cafes, and for various purpofes; fo as to be able to affure the Society that I have found them to anfwer.

PART I.

Concerning Undershot Water Wheels,

PLATE IV. Fig. 1, is a perfpective view of the machine for experiments on water wheels; wherein

A B C D is the lower ciftern, or magazine, for receiving the water, after it has quitted the wheel; and for fupplying

D E the upper ciftern, or head; wherein the water being raifed to any height required, by a pump, that height is fhewn by

F G, a fmall rod, divided into inches and parts; with a float at the bottom, to move the rod up and down, as the furface of the water rifes and falls.

H I

H I is a rod by which the fluice is drawn, and ftopped at any height required, by means of

K a pin, or peg, which fits feveral holes, placed in the manner of a diagonal fcale, upon the face of the rod H I.

G L is the upper part of the rod of the pump, for drawing the water out of the lower ciftern, in order to raife and keep up the furface thereof at its defired height, in the head D E; thereby to fupply the water expended by the aperture of the fluice.

M M is the arch and handle for working the pump, which is limited in its ftroke by

N, a piece for ftopping the handle from raifing the pifton too high; that alfo being prevented from going too low, by meeting the bottom of the barrel.

O is the cylinder, upon which a cord winds, and which, being conducted over the pullies P and Q, raifes

R, the fcale, into which the weights are put, for trying the power of the water.

S T, the two ftandards, which fupport the wheel, are made to flide up and down, in order to adjuft the wheel, as near as poffible to the floor of the conduit.

W the beam which fupports the fcale and pullies; this is reprefented as but little higher than the machine, for the fake of bringing the figure into a moderate compafs, but in reality is placed fifteen or fixteen feet higher than the wheel.

PLATE V. Fig. 2, is a fection of the fame machine, wherein the fame parts are marked with the fame letters as in fig. 1. Befides which

X X is the pump barrel, being five inches diameter, and eleven inches long.

Y is the pifton; and

Z the fixed valve.

G V is a cylinder of wood, fixed upon the pump-rod, and reaches above the furface of the water: this piece of wood being of fuch a thicknefs, that its fection is half the

area

Plate 5. page 28.

Scale of Inches to fig. 3.

Fig. 3.

Fig. 2.d

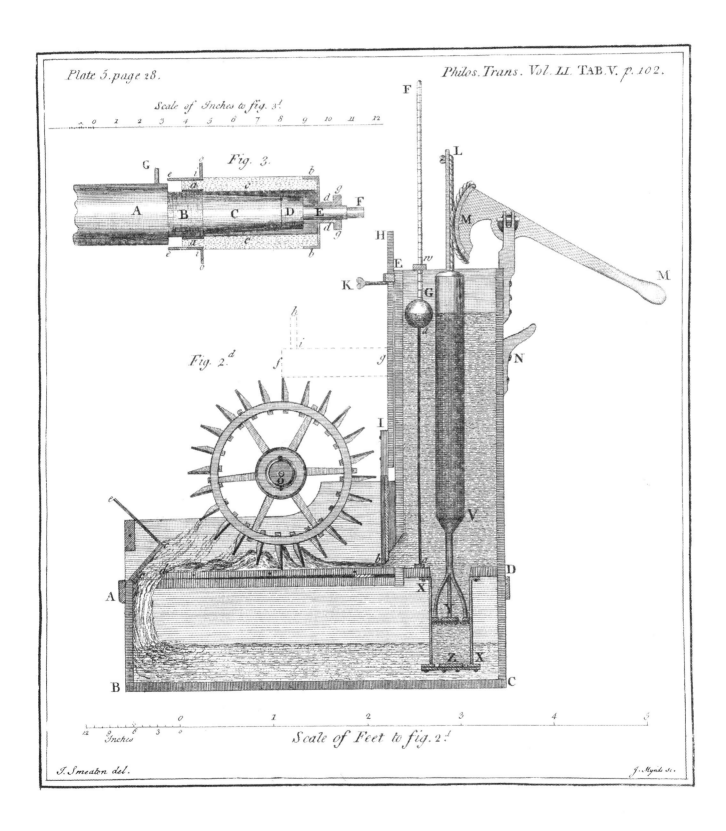

Scale of Feet to fig. 2.d

area of that of the pump-barrel, will caufe the furface of the water to rife in the head, as much while the pifton is defcending, as while it is rifing: and will thereby keep the gauge-rod F G more equally to its height.—*Note*, The arch and handle M M is here reprefented on a different fide to what it is fhewn in the preceding figures, in order that its dimenfions may the better appear.

a a fhews one of the two wires which ferve as directors to the float, in order that the gauge rod F G may be kept perpendicular ; for the fame purpofe alfo ferves *w*, a piece of wood with a hole to receive the gauge-rod, and keep it upright.

b is the aperture of the fluice.

c c a kant-board, for throwing the water more directly down the opening *c d*, into the lower ciftern ; and

c e is a floping board, for bringing back the water that is thrown up by the floats of the wheel.

Fig. 3, reprefents one end of the main axis, with a fection of the moveable cylinder, marked O in the preceding figures.

A B C D is the end of the axis ; whereof the parts

B and D are covered with ferrules or hoops of brafs.

E is a cylinder of metal ; whereof the part marked

F is the pivot or gudgeon.

c c is the fection of an hollow cylinder of wood, the diameter of the interior part being fomewhat larger than the cylindrical ferrule B.

a a is the fection of a ferrule of brafs, driven into the end of the hollow cylinder, and which is adjufted to that marked B, fo as to flide freely thereupon, but with as little fhake as poffible.

b b, *d d*, *g g*, reprefent the fection of a brafs ferrule, plate, and focket, fixed upon the other end of the hollow cylinder ; the focket *d d* being adjufted to flide freely upon the cylinder E, in the fame manner as the ferrule *a a* flides upon the cylinder B: the outer of the focket at

g g is formed into a fort of button; by pufhing whereof, the hollow cylinder will move backwards and forwards, or turn round at pleafure upon the cylindrical parts of the axis B and E.

e e, i i, o o, reprefent the fection of a brafs ferrule, alfo fixed upon the hollow cylinder: the edge of this ferrule

e e is cut into teeth, in the manner of a *contrate* wheel; and the edge thereof

o o is cut in the manner of a ratchet.

Of confequence, when the plate *b d d b* is pufhed clofe to the ferrule D, the teeth of the ferrule *e e* will lay hold of

G, a pin fixed into the axis; by which means the hollow cylinder is made to turn along with the wheel and axis: but being drawn back by the button *g g*, the hollow cylinder is thereby difengaged from the pin G, and ceafes turning.—*Note*, The weight in the fcale is prevented from running back, by a catch that plays in and lays hold of the ratchet *o o*.

By this means the hollow cylinder, upon which the cord winds and raifes the weight, is put in action and difcharged therefrom inftantaneoufly, while the wheel is in motion: for, without fome contrivance of this kind, it would not be eafy to make this fort of experiments with any tolerable degree of exactnefs.

The ufe of the apparatus now defcribed will be rendered more intelligible, by giving a general idea of what I had in view; but as I fhall be obliged to make ufe of a term which has heretofore been the caufe of difputation, I think it neceffary to affign the fenfe in which I would be underftood to ufe it; and in which I apprehend it is ufed by practical *mechanics*.

The word *power*, as ufed in practical mechanics, I apprehend, to fignify the exertion of ftrength, gravitation, impulfe, or preffure, fo as to produce motion: and by means of ftrength, gravitation, impulfe, or preffure, compounded with motion, to be capable

of

of producing an effect: and that no effect is properly mechanical, but what requires such a kind of power to produce it.

The raifing of a weight, relative to the height to which it can be raifed in a given time, is the moft proper meafure of power; or, in other words, if the weight raifed is multiplied by the height to which it can be raifed in a given time, the product is the meafure of the power raifing it; and confequently, all thofe powers are equal, whofe products, made by fuch multiplication, are equal: for if a power can raife twice the weight to the fame height; or the fame weight to twice the height, in the fame time that another power can, the firft power is double the fecond: and if a power can raife half the weight to double the height; or double the weight to half the height; in the fame time that another can, thofe two powers are equal. But *note*, all this is to be underftood in cafe of flow or equable motion of the body raifed; for in quick, accelerated, or retarded motions, the *vis inertiæ* of the matter moved will make a variation.

In comparing the effects produced by water-wheels with the powers producing them; or, in other words, to know what part of the original power is neceffarily loft in the application, we muft previoufly know how much of the power is fpent in overcoming the friction of the machinery, and the refiftance of the air; alfo what is the real velocity of the water at the inftant that it ftrikes the wheel; and the real quantity of water expended in a given time.

From the velocity of the water, at the inftant that it ftrikes the wheel, given, the height of head productive of fuch velocity can be deduced, from acknowledged and experimented principles of hydroftatics: fo that by multiplying the quantity, or weight of water, really expended in a given time, by the height of a head fo obtained; which muft be confidered as the height from which that weight of water had defcended in that given time; we fhall have a product, equal to the original power of the water, and clear of all uncertainty that would arife from the friction of the water, in paffing fmall apertures; and from all doubts, arifing from the different meafure of fpouting waters, affigned by different authors. On the other hand, the fum of the weights raifed by the action of this water, and of the weight required to overcome the friction and refiftance of the machine, multiplied by the height to which the weight can be raifed in the time given, the product will be equal to the effect of that power; and the proportion of the two products will be the proportion of the *power* to the *effect*. fo that by loading the wheel with different weights fucceffively, we fhall be able to determine at what particular load, and velocity of the wheel, the effect is a *maximum*.

The

The manner of finding the real velocity of the water, at the inftant of its ftriking the wheel; the manner of finding the value of the friction, refiftance, &c. in any given cafe; and the manner of finding the real expence of water, fo far as concerns the following experiments, without having recourfe to theory; being matters upon which the following determinations depend, it will be neceffary to explain them.

To determine the Velocity of the Water striking the Wheel.

It has already been mentioned, in the references to the figures, that weights are raifed by a cord winding round a cylindrical part of the axis. Firft, then, let the wheel be put in motion by the water, but without any weights in the fcale; and let the number of turns in a minute be fixty: now it is evident, that were the wheel free from friction and refiftance, that fixty times the circumference of the wheel would be the fpace through which the water would have moved in a minute; with that velocity wherewith it ftruck the wheel: but the wheel being incumbered by friction and re-fiftance, and yet moving fixty turns in a minute, it is plain that the velocity of the water muft have been greater than fixty circumferences before it met with the wheel. Let now the cord be wound round the cylinder, but contrary to the ufual way, and put a weight in the fcale; the weight fo difpofed (which may be called the *counter-weight*) will endeavour to affift the wheel in turning the fame way, as it would have been turned by the water: put therefore as much weight into the fcale as, without any water, will caufe it to turn fomewhat fafter than at the rate of fixty turns in a minute; fuppofe fixty-three; let it now be tried again by the water, affifted by the weight; the wheel therefore will now make more than fixty turns; fuppofe fixty-four: hence we conclude the water ftill exerts fome power in giving motion to rhe wheel. Let the weight be again increafed, fo as to make fixty-four and a half turns in a minute with-out water: let it once more be tried with water as before; and fuppofe it now to make the fame number of turns with water as without, viz. fixty-four and a half: hence it is evident, that in this cafe the wheel makes the fame number of turns in a minute, as it would do if the wheel had no friction or refiftance at all; becaufe the weight is equivalent thereto; for were it too little the water would accelerate the wheel beyond the weight; and if too great, retard it; fo that the water now becomes a *regulator* of the wheel's motion; and the velocity of its circumference becomes a meafure of the velocity of the water.

In like manner, in feeking the greateft product, or *maximum*, of effect; having found by trials what weight gives the greateft product, by fimply multiplying the

weight

weight in the scale by the number of turns of the wheel, find what weight in the scale, when the cord is on the contrary side of the cylinder, will cause the wheel to make the same number of turns the same way, without water; it is evident that this weight will be nearly equal to all friction and resistance taken together; and consequently, that the weight *in* the scale, with twice * the weight *of* the scale, added to the back or counter-weight, will be equal to the weight that could have been raised, supposing the machine had been without friction or resistance; and which multiplied by the height to which it was raised, the product will be the greatest effect of that power.

The Quantity of Water expended is found thus :

The pump made use of for replenishing the head with water was so carefully made, that, no water escaping back by the leathers, it delivered the same quantity of water at every stroke, whether worked quick or slow; and as the length of the stroke was limited, consequently the value of one stroke (or, on account of more exactness, twelve strokes) was known, by the height to which the water was thereby raised in the head; which, being of a regular figure, was easily measured. The sluice, by which the water was drawn upon the wheel, was made to stop at certain heights by a peg; so that when the peg was in the same hole, the aperture for the effluent water was the same. Hence the quantity of water expended by any given head, and opening of the sluice, may be obtained: for, by observing how many strokes a minute was sufficient to keep up the surface of the water at the given height, and multiplying the number of strokes by the value of each, the water expended by any given aperture and head in a given time will be given.

These things will be further illustrated by going over the *calculus* of one set of experiments.

Specimen of a Set of Experiments.

The sluice drawn to the first hole.

The water above the floor of the sluice - - - - - 30 inches.
Strokes of the pump in a minute - - - - - - 39½
The head raised by 12 strokes - - - - - - - 21
The wheel raised the empty scale, and made turns in a minute - - 80
With a counter-weight of 1 lb. 8 oz. it made - - - - - - - 85
Ditto tried with water - - - - - - - - - - - - - - 86

* The weight of the scale makes part of the weight both ways.

F   No

No.	Weight.	Turns in a minute.	Product.
	lb. oz.		
1	4 0	45	180
2	5 0	42	210
3	6 0	36½	217½
4	7 0	33¾	236¼
5	8 0	30	240 maximum.
6	9 0	26½	238½
7	10 0	22	220
8	11 0	16½	181¼
9	12	* ceafed working.	

Counter-weight, for 30 turns without water, 2 oz. in the fcale.

N. B. The area of the head was 105,8 fquare inches.
Weight of the empty fcale and pulley, 10 oz.
Circumference of the cylinder, 9 inches.
Circumference of the water-wheel, 75 ditto.

Reduction of the above Set of Experiments.

The circumference of the wheel, 75 inches, multiplied by 86 turns, gives 6450 inches for the velocity of the water in a minute; $\frac{1}{60}$ of which will be the velocity in a fecond, equal to 107,5 inches, or 8,96 feet, which is due to a head of fifteen inches †; and this we call the *virtual* or *effective* head.

The area of the head being 105,8 inches, this multiplied by the weight of water of the inch cubic, equal to the decimal ,579 of the ounce avoirdupoife, gives 61,26 ounces for the weight of as much water, as is contained in the head, upon one inch in depth, $\frac{1}{16}$ of which is 3,83 pounds; this multiplied by the depth 21 inches, gives 80,43lb. for the value of 12 ftrokes; and by proportion, 39½ (the number made in a minute) will give 264,7 lb. the weight of water expended in a minute.

---

* *N. B.* When the wheel moves fo flow as not to rid the water fo faft as fupplied by the fluice, the accumulated water falls back upon the aperture, and the wheel immediately ceafes moving.

† This is determined upon the common maxim of hydroftatics, that the velocity of fpouting waters is equal to the velocity that an heavy body would acquire in falling from the height of the refervoir; and is proved by the rifing of jets to the height of their refervoirs nearly.

Now,

Now, as 264,7 lb. of water may be confidered as having defcended through a fpace of 15 inches in a minute, the product of thefe two numbers 3970 will exprefs the *power* of the water to produce mechanical effects; which were as follows:

The velocity of the wheel at the *maximum*, as appears above, was 30 turns a minute; which multiplied by 9 inches, the circumference of the cylinder, makes 270 inches; but as the fcale was hung by a pulley and double line, the weight was only raifed half of this, viz. 135 inches.

The weight in the fcale at the *maximum* 8*lb*. 0 *oz*.
Weight of the fcale and pulley -   - 0 10
Counter-weight, fcale, and pulley - - 0 12
_____
Sum of the refiftance -   -   - 9 6
or *lb*. 9,375.

Now, as 9,375 lb is raifed 135 inches, thefe two numbers being multiplied together, the product is 1266, which expreffes the effect produced at a *maximum*: fo that the proportion of the *power* to the *effect* is as 3970 : 1266, or as 10 : 3,18.

But though this is the greateft *single* effect producible from the power mentioned, by the impulfe of the water upon an undershot wheel; yet, as the whole power of the water is not exhaufted thereby, this will not be the true ratio between the *power* of the water, and the *sum* of all the *effects* producible therefrom : for as the water muft neceffarily leave the wheel with a velocity equal to the wheel's circumference, it is plain that fome part of the power of the water muft remain after the quitting the wheel.

The velocity of the wheel at the *maximum* is 30 turns a minute; and confequently its circumference moves at the rate of 3,123 feet a fecond, which anfwers to a head 1,82 inches; this being multiplied by the expence of water in a minute, viz. 264,7 lb. produces 481 for the power *remaining* in the water after it has paffed the wheel: this being therefore deducted from the original power 3970, leaves 3489, which is that *part* of the power which is fpent in producing the effect 1266; and confequently the part of the power fpent in producing the effect, is to the greateft effect producible thereby as 3489 : 1266 : : 10 : 3,62, or as 11 to 4.

The *velocity of the water* ftriking the wheel has been determined to be equal to 86 circumferences of the wheel *per* minute, and the *velocity of the wheel* at the *maximum* to be 30; the velocity of the water will therefore be to that of the wheel as 86 to 30; or as 10 to 3,5, or as 20 to 7.

The

The *load at the maximum* has been fhown to be equal to 9 lb. 6 oz. and that the wheel ceafed moving with 12 lb in the fcale: to which, if the weight of the fcale is added, viz. 10 ounces *, the proportion will be nearly as 3 to 4 between the load at the *maximum* and *that* by which the wheel is ftopped.

It is fomewhat remarkable, that though the velocity of the wheel in relation to the water turns out greater than ⅓ of the velocity of the water, yet the impulfe of the water in the cafe of a *maximum* is more than double of what is affigned by theory; that is, inftead of ⅓ of the column, it is nearly equal to the whole column.

It muft be remembered, therefore, that in the prefent cafe, the wheel was not placed in an open river, where the natural current, after it has communicated its impulfe to the float, has room on all fides to efcape, as the theory fuppofes; but in a conduit, or race, to which the float being adapted, the water cannot otherwife efcape than by moving along with the wheel. It is obfervable, that a wheel working in this manner as foon as the water meets the float receiving a fudden check, it rifes up againft the float, like a wave againft a fixed object; infomuch that when the fheet of water is not a quarter of an inch thick before it meets the float, yet this fheet will act upon the whole furface of a float, whofe height is three inches; and confequently were the float no higher than the thicknefs of the fheet of water, as the theory alfo fuppofes, a great part of the force would have been loft, by the water dafhing over the float †.

---

* The refiftance of the air in this cafe ceafes, and the friction is not added, as 12 lb. in the fcale was fufficient to ftop the wheel after it had been in full motion; and therefore fomewhat more than a counterbalance to the impulfe of the water.

† Since the above was written I find that Professor Euler, in the Berlin Acts for the year 1748, in a memoir entitled *Maxims pour aranger le plus avantageusment les machines destinees à elever de l'eau par le moyen de pompes*, page 192. § 9. has the following passage; which feems to be the more remarkable, as I do not find he has given any demonftration of the principle therein contained, either from theory or experiment; or has made any ufe thereof in his calculations on this fubject —" Cependant dans ce cas " puifque l'eau eft reflechie, & qu'elle decoule fur les aubes yers les cotes, elle y exerce encore une force " particuliere, dont l'effet de l'impulfion fera augmenté; & experience jointe a la theorie a fait voir que " dans ce cas, la force eft prefque double: de forte qui'il faut prendre le double de le fection du fil d'eau " pour ce qui repond dans ce cas-a le furface des aubes, pourvu qu'elles foient affez larges pour recevoir " ce fupplement de force. Car fi les aubes nétoient plus larges que le fil, on trait d'eau on ne devroit " prendre que ne fimple fection, tout comme dans le premier cas, ou l'aube toute entire eft pappee " par l'eau."

TABLE

TABLE I.

Number.	Height of water in the cistern.	Turns of the wheel un loaded.	Virtual head deduced therefrom.	Turns at the maximum	Load at the equilibrium.		Load at the maximum.		Water expended in a minute.	POWER.	EFFECT.	Ratio of the power and effect.	Ratio of the velocity of the water and wheel.	Ratio of the load at the equilibrium, to the load at the maximum.	Experiments.
	In.		In.		lb.	oz	lb.	oz.							
1	33	88	15,85	30,	13	10	10	9	275,	4358	1411	10:3,24	10:3,4	10:7,75	
2	30	86	15,0	30,	12	10	9	6	264,7	3970	1266	10:3,2	10:3,5	10:7,4	
3	27	82	13,7	28,	11	2	8	6	243,	3329	1044	10:3,15	10:3,4	10:7,5	
4	24	78	12,3	27,7	9	10	7	5	235,	2890	901,4	10:3,12	10:3,55	10:7,53	At
5	21	75	11,4	25,9	8	10	6	5	214,	2439	735.7	10:3,02	10:3,45	10:7,32	the
6	18	70	9,95	23,5	6	10	5	5	199,	1970	561,8	10:2,85	10:3,36	10:8,02	1st
7	15	65	8,54	23,4	5	2	4	4	178,5	1524	442,5	10:2,9	10:3,6	10:8,3	hole.
8	12	60	7,29	22,	3	10	3	5	161,	1173	328	10:2,8	10:3,77	10:9,1	
9	9	52	5,47	19,	2	12	2	8	134,	733	213,7	10:2,9	10:3.65	10:9,1	
10	6	42	3,55	16,	1	12	1	10	114,	404,7	117	10:2,82	10:3,8	10:9,3	
11	24	84	14,2	30,75	13	10	10	14	342,	4890	1505	10:3,075	10:3,66	10:7,9	
12	21	81	13,5	29,	11	10	9	6	297,	4009	1223	10:3,01	10:3,62	10:8,05	
13	18	72	10,5	26,	9	10	8	7	285,	2993	975	10:3,25	10:3,6	10:8,75	At
14	15	69	9,6	25,	7	10	6	14	277,	2659	774	10:2,92	10:3,62	10:9,	the
15	12	63	8,0	25,	5	10	4	14	234,	1872	549	10:2,94	10:3,97	10:8,7	2d.
16	9	56	6,37	23,	4	0	3	13	201,	1280	390	10:3,05	10:4,1	10:9,5	
17	6	46	4,25	21,	2	8	2	4	167,5	712	212	10:2,98	10:4,55	10:9,	
18	15	72	10.5	29,	11	10	9	6	357,	3748	1210	10:3,23	10:4,02	10:8,05	
19	12	66	8,75	26,75	8	10	7	6	330,	2887	878	10:3,05	10:4,05	10:8,1	The
20	9	58	6,8	24.5	5	8	5	0	255,	1734	541	10:3,01	10:4,22	10:9,1	3d.
21	6	48	4,7	23,5	3	2	3	0	228,	1064	317	10:2,99	10:4,9	10:9,6	
22	12	68	9,3	27,	9	2	8	6	359,	3338	1006	10:3,02	10:3,97	10:9,17	
23	9	58	6 8	26,25	6	2	5	13	332,	2257	686	10:3,04	10:4,52	10:9,5	4th
24	6	48	4,7	24,5	3	12	3	8	262,	1231	385	10:3,13	10:5,1	10:9,35	
25	9	60	7,29	27,3	6	12	6	6	355,	2588	783	10:3,03	10 4,55	10:9,45	5th.
26	6	50	5,03	24,6	4	6	4	1	307,	1544	450	10:2,92	10 4,9	10:9,3	
27	6	50	5,03	26,	4	15	4	9	360,	1811	534	10:2,95	10:5,2	10:9,25	6th.
1.	2	3	4.	5	6.		7.		8.	9.	10.	11.	12.	13	

In further confirmation of what is already delivered, I have adjoined a table, (TABLE I.) containing the refult of 27 fets of experiments, made and reduced in the manner above fpecified. What remains of the theory of underfhot wheels, will naturally follow from a comparifon of the different experiments together.

Maxims and Obfervations deduced from the foregoing Table of Experiments.

Maxim I. That the virtual or effective head being the fame, the effect will be nearly as the quantity of water expended.

This will appear by comparing the contents of the columns 4, 8, and 10, in the foregoing fets of experiments; as for

Example 1ft, taken from No. 8, and 25, viz.

No.   Virtual Head. Water expended.   Effect.
8 - - - - 7,29 - - - 161 - - - - - 328
25 - - - - 7,29 - - - 355 - - - - - 785

Now the heads being equal, if the effects are proportioned to the water expended, we fhall have by maxim 1ft, 161 : 355 :: 328 : 723; but 723 falls fhort of 785, as it turns out in experiment, according to No. 25, by 62; the effect therefore of No. 25, compared with No. 8, is greater than according to the prefent maxim in the ratio of 14 to 13.

The foregoing example, with four fimilar ones, are feen at one view in the following Table.

Examples.	No. Tab. I.	Virtual Head.	Expence of Water.	Effect.	Comparifon.	Variation.	Proportional Variation.
		Inch	*lb.*				
1ft	8 25	7,29 7,29	161 355	328 785	161 : 355 :: 328 : 723	62 +	14 : 13
2d	13 18	10,5 10,5	285 357	975 1210	285 : 357 :: 975 : 1221	11 —	121 : 122
3d	22 23	6,8 6,8	255 332	541 686	255 : 332 :: 541 : 704	18 —	38 : 39
4th	21 24	4,7 4,7	228 262	317 385	228 : 262 :: 317 : 364	21 +	18 : 17
5th	26 27	5,03 5,03	307 360	450 534	307 : 360 :: 450 : 531	3 +	178 : 177

Hence therefore, in comparing different experiments, as fome fall fhort, and others exceed the *maximum*, and all agree therewith, as near as can be expected, in an affair where fo many different circumftances are concerned, we may, according to the laws of reafoning by induction, conclude the maxim true ; viz. that the effects are nearly as the quantity of water expended.

Maxim II. That the expence of water being the fame, the effect will be nearly as the height of the virtual or effective head.

This alfo will appear by comparing the contents of columns 4, 8, and 10, in any of the fets of experiments.

Example 1ft, of No. 2, anu No. 24, viz.

No.	Virtual Head.	Expence.	Effect.
2	15	264,7	1266
24	4,7	262	385

Now, as the expences are not quite equal, we muft proportion one of the effects accordingly : thus

$$\text{by maxim 1ft, } 262 : 264,7 :: 385 : 389$$
$$\text{and by max. 2d, } 15 : 4,7 :: 1266 : 397$$
$$\text{Difference} - - - - 8$$

The effect therefore of No. 24, compared with No. 2, is lefs than according to the prefent maxim in the ratio of 49 : 50.

The foregoing, and two other fimilar examples, are comprifed in the following Table :

Examples.	No. Table I.	Virtual Head.	Expence of Water.	Effect.	Comparifon.	Variation.	Proportional Variation.
1ft {	2 24	15 4,7	264,7 262	1266 385	Max. 1ft, 262 : 264,7 :: 385 : 319 Max. 2d, 15 : 4,7 :: 1266 : 397	8 —	49 : 50
2d	1 10	15,85 3,55	275 114	1411 117	Max. 1ft, 114 : 275 :: 117 : 282 Max. 2d, 15,85 : 3,55 :: 1411 : 316	34 —	8 : 9
3d {	11 17	14,2 4,25	342 1675	1505 212	Max. 1ft, 167,5 : 342 :: 212 : 433 Max. 2d, 14,2 : 4,25 :: 1505 : 450	17 —	25 : 26

Maxim III. That the quantity of water expended being the fame, the effect is nearly as the fquare of its velocity.

This will appear by comparing the contents of columns 3, 8, and 10, in any of the fets of experiments; as for

Example 1ft, of No. 2, with No. 24, viz.

No.	Turns in a minute.	Expence.	Effect.
2 - - - -	86 - - - - - -	264,7 - - -	1266
24 - - - -	48 - - - - - -	262 - - -	385

The velocity being as the number of turns, we fhall have,

by maxim 1ft, $262 : 264,7 : : 385 : 389$

and by maxim 3d. $\begin{Bmatrix} 86^2 : 48^2 \\ 7396 : 2304 \end{Bmatrix} : : 1266 : 394$

Difference - - - - 5

The effect therefore of No. 24, compared with No. 2, is lefs than by the prefent maxim in the ratio of 78 : 79.

The foregoing, and three other fimilar examples, are comprifed in the following Table:

Examples.	No. Table I.	Turns in a Minute.	Expence of Water.	Effect.	Comparifon.		Variation.	Proportional Variation.
1ft	2	86	264,7	1266	Max. 1ft, $262:264,7 : : 385 : 389$		5 —	78 : 79
	24	48	262	385	Max. 3d, $\begin{Bmatrix} 86^2 : 48^2 \\ 7396 : 2304 \end{Bmatrix} : : 1266 : 394$			
2	1	88	275	1411	Max. 1ft, $114 : 275 : : 117 : 282$		39 —	7 : 8
	10	42	114	117	Max. 3d, $\begin{Bmatrix} 88^2 : 42^2 \\ 7744 : 1764 \end{Bmatrix} : : 1411 : 321$			
3d	11	84	342	1505	Max. 1ft, $167,5 : 342 : : 212 : 433$		18 —	24 : 25
	17	46	167,5	212	Max. 3d, $\begin{Bmatrix} 84^2 : 46^2 \\ 7056 : 2116 \end{Bmatrix} : : 1505 : 451$			
4th	18	72	357	1210	Max. 1ft, $228 : 357 : : 317 : 496$		42 —	12 : 13
	21	48	228	317	Max. 3d, $\begin{Bmatrix} 72^2 : 48^2 \\ 5184 : 2301 \end{Bmatrix} : : 1210 : 538$			

Maxim IV. The aperture being the fame, the effect will be nearly as the cube of the velocity of the water.

This alfo will appear by comparing the contents of columns 3, 8, and 10; as for

Example 1ft, of No. 1, and No. 10, viz.

No.	Turns.	Expence.	Effect.
1	88	275	1411
10	42	114	117

Lemma. It muft here be obferved, that if water paffes out of an aperture, in the fame fection, but with different velocities, the expence will be proportional to the velocity; and therefore converfely, if the expence is not proportional to the velocity, the fection of the water is not the fame.

Now, comparing the water difcharged with the turns of No. 1, and 10, we fhall have $88 : 42 :: 275 : 131,2$; but the water difcharged by No. 10, is only 114 lb. therefore, though the fluice was drawn to the fame height in No. 10, as in No. 1, yet the fection of the water paffing out, was lefs in No. 10, than No. 1, in the proportion of 114 to 131,2; confequently, had the effective aperture or fection of the water been the fame in No. 10, as in No. 1, fo that 131,2 lb. of water had been difcharged inftead of 114, the effect would have been increafed in the fame proportion; that is

$$\text{by the } \textit{Lemma,} \quad 88 : \quad 42 \quad :: \quad 275 : \quad 131,2$$
$$\text{by maxim 1ft,} \quad 114 : 131,2 \quad :: \quad 117 : \quad 134,5$$
$$\text{and by maxim 4th.} \left\{ \begin{matrix} 83_3 & : & 42_3 \\ 681472 & : & 74088 \end{matrix} \right\} :: 1411 : \quad 153,5$$

$$\text{Difference} \quad - \quad - \quad - \quad - \quad 19$$

The effect therefore of No. 10, compared with No. 1, is lefs than it ought to be by the prefent maxim in the ratio of $7 : 8$.

The

The foregoing, and three other fimilar examples, are contained in the following Table :

Examples.	No. Table I.	Turns in a Minute.	Expence of Water.	Effect.	Comparifon.	Variation.	Proportional Variation.
1ft	{ 1 / 10	88 / 42	275 / 114	1411 / 117	Lemma. 88 : 42 :: 275 : 131,2 Max. 1. 114 : 131,2 :: 117 : 134,5 Max. 4. 88³ : 42³ :: 1411 : 153,5	19—	7 : 8
2d	{ 11 / 17	84 / 46	342 / 167,5	1505 / 212	Lemma. 84 : 46 :: 342 : 187,3 Max.1. 167,5 : 187,3 :: 212 : 237 Max. 4. 84³ : 46³ :: 1505 : 247	10—	23 : 24
3d	{ 18 / 21	72 / 48	357 / 228	1210 / 317	Lemma. 72 : 48 :: 357 : 238 Max. 1. 228 : 238 :: 317 : 331 Max. 4. 72³ : 48³ :: 1210 : 355	24—	14 : 15
4th	{ 22 / 24	68 / 48	359 / 262	1006 / 385	Lemma. 68 : 48 :: 359 : 253,4 Max. 1. 262 : 253,4 :: 385 : 372 Max. 4. 68³ : 48³ :: 1006 : 354	18+	20 : 19

Observations.

Obferv. 1ft. On comparing column 2d and. 4th, Table I. it is evident that the *virtual head* bears no certain proportion to the *head of water;* but that when the aperture is greater, or the velocity of the water iffuing therefrom lefs, they approach nearer to a coincidence; and confequently in the large openings of mills and fluices, where great quantities of water are difcharged from moderate heads, the head of water, and virtual head determined from the velocity, will nearly agree, as experience confirms.

Obferv. 2d. Upon comparing the feveral proportions between the *power* and *effect* in column 11th, the moft general is that of 10 to 3; the extremes 10 to 3,2 and 10 to 2,8; but as it is obfervable, that where the quantity of water, or the velocity thereof; that is, where the power is greateft, the 2d term of the ratio is greateft alfo: we may therefore well allow the proportion fubfifting in large works, as 3 to 1.

Obferv.

Observ. 3d. The proportions of *velocities* between the *water* and *wheel* in column 12, are contained in the limits of 3 to 1 and 2 to 1; but as the greater velocities approach the limit of 3 to 1, and the greater quantity of water approaches to that of 2 to 1, the beſt general proportion will be that of 5 to 2.

Observ. 4th. On comparing the numbers in column 13, it appears, that there is no certain ratio between the *load* that the wheel will carry at its *maximum*, and what will totally ſtop it; but that they are contained within the limits of 20 to 19, and of 20 to 15; but as the effect approaches neareſt to the ratio of 20 to 15, or of 4 to 3, when the power is greateſt, whether by increaſe of velocity, or quantity of water, this ſeems to be the moſt applicable to large works; but as the load that a wheel ought to have, in order to work to the beſt advantage, can be aſſigned, by knowing the effect it ought to produce, and the velocity it ought to have in producing it; the exact knowledge of the greateſt load it will bear, is of the leſs conſequence in practice.

It is to be noted, that in all the examples under the laſt three of the four preceding maxims, the effect of the leſſer power falls ſhort of its due proportion to the greater, when compared by its maxim; except the laſt example of maxim fourth: and hence, if the experiments are taken ſtrictly, we muſt infer, that the effects increaſe and diminiſh in an higher ratio than thoſe maxims ſuppoſe; but as the deviation is not very conſiderable, the greateſt being about one-eighth of the quantity in queſtion; and as it is not eaſy to make experiments of ſo compounded a nature with abſolute preciſion; we may rather ſuppoſe, that the leſſer power is attended with ſome friction, or works under ſome diſadvantage, which has not been duly accounted for; and therefore we may conclude, that theſe maxims will hold very nearly, when applied to works in large.

After the experiments above mentioned were tried, the wheel, which had originally twenty-four floats, was reduced to twelve; which cauſed a diminution in the effect, on account of a greater quantity of water eſcaping between the floats and the floor; but a circular ſweep being adapted thereto, of ſuch a length, that one float entered the curve before the preceding one quitted it, the effect came ſo near to the former, as not to give hopes of advancing it by increaſing the number of floats beyond twenty four in this particular wheel.

PART

PART II.

Concerning Overshot Wheels.

Read before the Royal Society, May 24th, 1759.

IN the former part of this effay, we have confidered the impulfe of a confined ftream, acting on *Undershot Wheels.* We now proceed to examine the power and application of water, when acting by its *gravity* on *Overshot Wheels.*

In reafoning without experiment, one might be led to imagine, that however different the mode of application is; yet, that whenever the fame quantity of water defcends through the fame perpendicular fpace, that the natural effective power would be equal; fuppofing the machinery free from friction, equally calculated to receive the full effect of the power, and to make the moft of it: for if we fuppofe the height of a column of water to be 30 inches, and refting upon a bafe or aperture of one inch fquare, every cubic inch of water that departs therefrom will acquire the fame velocity, or *momentum*, from the uniform preffure of 30 cubic inches above it, that one cubic inch let fall from the top will acquire in falling down to the level of the aperture; viz. fuch a velocity as, in a contrary direction, would carry it to the level from whence it fell;* one would therefore fuppofe, that a cubic inch of water, let fall through a fpace of 30 inches, and there impinging upon another body, would be capable of producing an equal effect by collifion, as if the fame cubic inch had defcended through the fame fpace with a flower motion, and produced its effects gradually: for in both cafes gravity acts upon an equal quantity of matter, through an equal fpace; † and confequently, that whatever was the ratio between the power and effect in underfhot wheels, the fame would obtain in overfhot, and indeed in all others: yet, however conclufive this reafoning may feem, it will appear, in the courfe of the following deductions, that the effect of the gravity of defcending bodies is very different from

---

* This a confequence of the rifing of jets to the height of their refervoirs nearly.

† Gravity, it is true, acts a longer fpace of time upon the body that defcends flowly than upon that which falls quickly; but this cannot occafion the difference in the effect: for an elaftic body falling through the fame space in the fame time, will, by collifion upon another elaftic body, rebound nearly to the height from which it fell or, by communicating its motion, caufe an equal one to afcend to the fame height.

the

the effect of the ftroke of fuch as are *non elaftic*, though generated by an equal mechanical power.

The alterations in the machinery already defcribed, to accommodate the fame for experiments on overfhot wheels, where principally as follows :

PLATE V. Fig. 2. The fluice I *b* being fhut down, the rod H I was unfcrewed and taken off.

The underfhot water-wheel was taken off the axis, and inftead thereof an overfhot wheel of the fame diameter was put into its place. *Note*, This wheel was two inches in the fhroud or depth of the bucket ; the number of the buckets was 36.

The ftandards S and T, PLATE IV. were raifed half an inch, fo that the bottom of the wheel might be clear of ftagnant water.

A trunk, for bringing the water upon the wheel, was fixed according to the dotted lines *f g*, Fig. 2. The aperture was adjufted by a fhuttle *h i*, which alfo clofed up the outer end of the trunk, when the water was to be ftopped.

Fig. 3. The ratchet *o o*, not being of one piece of metal with the ferrule *e e, i i* (though fo defcribed before, to prevent unneceffary diftinctions) was with its catch turned the contrary fide ; confequently the moveable barrel would do its office equally, notwithftanding the water-wheel, when at work, moved the contrary way.

Specimen

Specimen of a Set of Experiments.

Head 6 inches.

14 $\frac{1}{2}$ ftrokes of the pump in a minute, 12 ditto = 80 lb. *

Weight of the fcale (being wet) 10$\frac{1}{2}$ oz.

Counterweight for 20 turns, befides the fcale, 3 oz.

No.	Weight in the Scale.	Turns.	Product.	Obfervations.
1	- - - 0 *lb.* - -	- 60 -	- - -	Threw moft part of the water out of the wheel.
2	- - 1 - -	- 56 -	- - -	
3	- - 2 - -	- 52 -	- - -	
4	- - 3 - -	- 49 -	- 147	Received the water more quietly.
5	- - 4 - -	- 47 -	- 188	
6	- - 5 - -	- 45 -	- 225	
7	- - 6 - -	- 42$\frac{1}{2}$ -	- 255	
8	- - 7 - -	- 41 -	- 287	
9	- - 8 - -	- 38$\frac{1}{2}$ -	- 308	
10	- - 9 - -	- 36$\frac{1}{2}$ -	- 328$\frac{1}{2}$	
11	- - 10 - -	- 35$\frac{1}{2}$ -	- 355	
12	- - 11 - -	- 32$\frac{3}{4}$ -	- 360$\frac{1}{2}$	
13	- - 12 - -	- 31$\frac{1}{4}$ -	- 375	
14	- - 13 - -	- 28$\frac{1}{8}$ -	- 370$\frac{1}{2}$	
15	- - 14 - -	- 27$\frac{1}{2}$ -	- 385	
16	- - 15 - -	- 26 -	- 390	
17	- - 16 - -	- 24$\frac{1}{2}$ -	- 392	
18	- - 17 - -	- 22$\frac{3}{4}$ -	- 386$\frac{3}{4}$	
19	- - 18 - -	- 21$\frac{3}{4}$ -	- 391$\frac{1}{2}$	
20	- - 19 - -	- 20$\frac{3}{4}$ -	- 394$\frac{1}{4}$	maximum.
21	- - 20 - -	- 19$\frac{3}{4}$ -	- 395	
22	- - 21 - -	- 18$\frac{1}{4}$ -	- 388$\frac{1}{4}$	
23	- - 22 - -	- 18 -	- 396	Worked irregular.
24	- - 23 - -	- Overfet by its load.		

* The small difference, in the value of 12 strokes of the pump, from the former experiments, was owing to a small difference in the length of the stroke, occasioned by the warping of the wood.

Reduction

Reduction of the preceding Specimen.

In thefe experiments the head being 6 inches, and the height of the wheel 24 inches, the whole defcent will be 30 inches: the expence of water was $14\frac{1}{2}$ ftrokes of the pump in a minute, whereof 12 contained 80 lb.; therefore the water expended in a minute was $96\frac{1}{3}$ lb. which, multiplied by 30 inches, gives the *power* $= 2900$.

If we take the 20th experiment for the *maximum*, we fhall have $20\frac{3}{4}$ turns in a minute, each of which raifed the weight of $4\frac{1}{2}$ inches, that is, $93,37$ inches in a minute. The weight in the fcale was 19 lb. the weight of the fcale $10\frac{1}{2}$ oz.; the counterweight 3 oz. in the fcale, which, with the weight of the fcale $10\frac{1}{2}$ oz. makes in the whole $20\frac{1}{2}$ lb. which is the whole refiftance or load: this, multiplied by $93,37$ inches, makes 1914 for the effect.

The *ratio*, therefore of the *power* and *effect* will be as $2900 : 1914$, or as $10 : 6,6$, or as $3 : 2$ nearly.

But if we compute the power from the height of the wheel only, we fhall have $96\frac{1}{3}$ lb. multiplied by 24 inches $= 2320$ for the *power*, and this will be to the *effect* as $2320 : 1914$, or as $10 : 82$, or as $5 : 4$ nearly.

The reduction of this fpecimen is fet down in No. 9, of the following Table; and the reft were deduced from a fimilar fet of experiments, reduced in the fame manner.

TABLE

TABLE II.

Containing the Result of Sixteen Sets of Experiments on Overshot Wheels.

Number.	Whole descent.	Water expended in a minute.	Turns at the maximum in a min.	Weight raised at the maximum.	Power of the whole descent.	Power of the Wheel.	Effect.	Ratio of the whole power and effect.	Ratio of power of the wheel and effect.	Mean ratio.
	Inch.	lb.		lb.						
1	27	30	19	6½	810	720	556	10 : 6,9	10 : 7,7	Medium 10:8,1
2	27	56⅔	16¼	14½	1530	1360	1060	10 : 6,9	10 : 7,8	
3	27	56⅔	20¾	12¼	1530	1360	1167	10 : 7,6	10 : 8,4	
4	27	63½	20½	13½	1710	1524	1245	10 : 7,3	10 : 8,2	
5	27	76⅔	21½	15½	2070	1840	1500	10 : 7,3	10 : 8,2	
6	28½	73⅓	18¾	17½	2090	1764	1476	10 : 7,	10 : 8,4	10:8,2
7	28¼	96⅔	20¼	20½	2755	2320	1868	10 : 6,8	10 : 8,	
8	30	90	20	19½	2700	2160	1755	10 : 6,5	10 : 8,1	10:8,2
9	30	96⅔	20¾	20½	2900	2320	1914	10 : 6,6	10 : 8,2	
10	30	113⅓	21	23½	3400	2720	2221	10 : 6,5	10 : 8,2	
11	33	56⅔	20¼	13½	1870	1360	1230	10 : 6,6	10 : 9,	10:8,5
12	33	106⅔	22¼	21½	3520	2560	2153	10 : 6,1	10 : 8,4	
13	33	146⅔	23	27½	4840	3520	2846	10 : 5,9	10 : 8,1	
14	35	65	19¾	16½	2275	1560	1466	10 : 6,5	10 : 9,4	10:8,5
15	35	120	21½	25½	4200	2880	2467	10 : 5,9	10 : 8,6	
16	35	163½	25	26½	5728	3924	2981	10 : 5,2	10 : 7,6	
1.	2.	3.	4.	5.	6.	7.	8.	9.	10.	11.

Observations and Deductions from the foregoing Experiments.

I. Concerning the Ratio between the Power and Effect of Overshot Wheels.

The effective power of the water must be reckoned upon the whole descent; because it must be raised that height, in order to be in a condition of producing the same effect a second time.

The

The ratios between the *powers* fo eftimated, and the *effects* at the *maximum* deduced from the feveral fets of experiments, are exhibited at one view in column 9. of Table II.; and from hence it appears, that thofe ratios differ from that of 10 to 7,6 to that of 10 : 5,2, that is, nearly from 4 : 3 to 4 : 2. In thofe experiments where the heads of water and quantities expended are leaft, the proportion is nearly as 4 : 3 ; but where the heads and quantities are greateft, it approaches nearer to that of 4 : 2 ; and by a medium of the whole, the ratio is that of 3 : 2 nearly. We have feen before, in our obfervations upon the effects of underfhot wheels, that the general ratio of the power to the effect, when greateft was 3 : 1 ; *the effect therefore of overshot wheels, under the same circumstances of quantity and fall, is at a medium double to that of the undershot :* and, as a confequence thereof, *that non elastic bodies, when acting by their impulse or collision, communicate only a part of their original power ;* the other part being fpent in changing their figure in confequence of the ftroke.

The powers of water computed from the height of the wheel only, compared with the effects, as in column 10, appear to obferve a more conftant ratio : for if we take the medium of each clafs, which is fet down in column 11, we fhall find the extremes to differ no more than from the ratio of 10 : 8,1 to that of 10 : 8,5 ; and as the fecond term of the ratio gradually increafes from 8,1 to 8,5, by an increafe of head from 3 inches to 11, the excefs of 8,5 above 8,1 is to be imputed to the fuperior impulfe of the water at the head of 11 inches above that of 3 inches : fo that if we reduce 8,1 to 8, on account of the impulfe of the 3 inch head, *we shall have the ratio of the power, computed upon the height of the wheel only, to the effect at a maximum as* 10 : 8, *or as* 5 : 4 *nearly :* and from the quality of the ratio between power and effect fub-fifting where the conftructions are fimilar, we muft infer, *that the effects,* as well as the powers, *are as the quantities of water and perpendicular heights multiplied together respectively.*

II. Concerning the moft proper Height of the Wheel in Proportion to the whole Defcent.

We have already feen from the preceding obfervation, that the effect of the fame quantity of water, defcending through the fame perpendicular fpace, is double, when acting by its gravity upon an overfhot wheel, to what the fame produces when acting by its impulfe upon an underfhot. It alfo appears, that by increafing the head from 3 inches to 11, that is, the whole defcent, from 27 inches to 35, or in the ratio of 7 to 9 nearly, the effect is advanced no more than in the ratio of 8,1 to 8,4, that is, as 7 : 7,26 ; and

H   confequently

confequently the increafe of effect as not 1-7th of the increafe of perpendicular height. Hence it follows, *that the higher the wheel is in proportion to the whole descent, the greater will be the effect ;* becaufe it depends lefs upon the impulfe of the head, and more upon the gravity of the water in the buckets: and if we confider how obliquely the water iffuing from the head muft ftrike the buckets, we fhall not be at a lofs to account for the little advantage that arifes from the impulfe thereof: and fhall immediately fee of how little confequence this impulfe is to the effect of an overfhot wheel. However, as every thing has its limits, fo has this: for thus much is defirable *that the water should have somewhat greater velocity, than the circumference of the wheel, in coming thereon ;* otherwife the wheel will not only be retarded, by the buckets ftriking the water, but thereby dafhing a part of it over, fo much of the power is loft.

The velocity that the circumference of the wheel ought to have, being known by the following deductions, the head requifite to give the water its proper velocity is eafily computed from the common rules of hydroftatics; and will be found much lefs than what is generally practifed.

III. Concerning the Velocity of the Circumference of the Wheel, in order to produce the greateft Effect.

If a body is let fall freely from the furface of the head to the bottom of the defcent, it will take a certain time in falling; and in this cafe the whole action of gravity is fpent n giving the body a certain velocity: but if this body in falling is made to act upon fome other body, fo as to produce a mechanical effect, the falling body will be retarded; becaufe a part of the action of gravity is then fpent in producing the effect, and the remainder only giving motion to the falling body: and therefore *the slower a body descends, the greater will be the portion of the action of gravity applicable to the producing a mechanical effect ;* and in confequence the greater that effect may be.

If a ftream of water falls into the bucket of an overfhot wheel, it is there retained until the wheel by moving round difcharges it: of confequence the flower the wheel moves, the more water each bucket will receive: fo that what is loft in fpeed, is gained by the preffure of a greater quantity of water acting in the buckets at once: and if confidered only in this light, the mechanical power of an overfhot wheel to produce effects will be equal whether it moves quick or flow: but if we attend to what has been juft now obferved of the falling body, it will appear that fo much of the action of gravity, as is employed in giving the wheel and water therein a greater velocity, muſt be ſubtracted from its preſſure upon the buckets; ſo that, though the product made by multiplying the number of cubic inches of water acting in the wheel at once by its velocity will be the ſame in all caſes; yet, as each cubic inch, when the velocity is *greater* does not preſs ſo much upon the bucket as when it is *less*, the power of the water to produce effects will be greater in the leſs velocity than in the greater: but hence we are led to this general rule, *that* cæteris paribus, *the less the velocity of the wheel, the greater will be the effect thereof.* A confirmation of this doctrine, together with the limits it is ſubject to in practice, may be deduced from the foregoing ſpecimen of a ſet of experiments,

From theſe experiments it appears that, when the wheel made about 20 turns in a minute, the effect was near upon, the greateſt. When it made 30 turns, the effect was diminiſhed about $\frac{1}{20}$ part; but that when it made 40, it was diminiſhed about $\frac{1}{4}$; when it made leſs than $18\frac{1}{4}$, its motion was irregular; and when it was loaded ſo as not to admit its making 18 turns, the wheel was overpowered by its load.

It is an advantage in practice, that the velocity of the wheel ſhould not be diminiſhed further than what will procure ſome ſolid advantage in point of power: becauſe *cæteris paribus*, as the motion is ſlower, the buckets muſt be made larger; and the wheel being more loaded with water, the ſtreſs upon every part of the work will be increaſed in proportion: *the best velocity for practice therefore will be such, as when the wheel here used made about 30 turns in a minute; that is, when the velocity of the circumference is a little more than 3 feet in a second.*

Experience confirms, that this velocity of 3 feet in a ſecond is applicable to the higheſt overſhot wheels, as well as the loweſt; and all other parts of the work being properly adapted thereto, will produce very nearly the greateſt effect poſſible: however, this alſo is certain from experience, that *high wheels may deviate further from this rule, before they will lose their power, by a given aliquot part of the whole, than low ones can be admitted to do;* for a wheel of 24 feet high may move at the rate of ſix feet per ſecond without loſing any conſiderable part of its power*; and on the other hand, I have ſeen a wheel of 33 feet high, that has moved very ſteadily and well with a velocity but little exceeding 2 feet.

---

* The 24 feet wheel going at 6 feet in a ſecond, ſeems owing to the small proportion that the head (requiſite to give the water the proper velocity of the wheel) bears to the whole height.

IV. Concerning

IV. Concerning the Load for an Overſhot Wheel, in order that it may produce a Maximum.

The maximum load for an overshot wheel, is that which reduces the circum-ferences of the wheel to its proper velocity ; and this will be known, by dividing the effect it ought to produce in a given time by the ſpace intended to be deſcribed by the circumference of the wheel in the ſame time ; the quotient will be the reſiſtance overcome at the circumference of the wheel, and is equal to the load required, the fric-tion and reſiſtance of the machinery included.

V. Concerning the greateſt poſſible Velocity of an Overſhot Wheel.

The greateſt velocity that the circumference of an overſhot wheel is capable of, depends jointly upon the diameter or *height* of the wheel, and the velocity of falling bodies ; for it is plain that the velocity of the circumference can never be greater, than to deſcribe a ſemi-circumference, while a body let fall from the top of the wheel will deſcend through its diameter ; nor indeed quite ſo great, as a body deſcending through the ſame perpendicular ſpace cannot perform the ſame in ſo ſmall a time when paſſing through a ſemi-circle, as would be done in a perpendicular line. Thus, if a wheel is 16 feet 1 inch high, a body will fall through the diameter in one ſecond : this wheel therefore can never arrive at a velocity equal to the making one turn in two ſeconds ; but, in reality, an overſhot wheel can never come near this velocity ; for when it acquires a certain ſpeed, the greateſt part of the water is prevented from entering the buckets ; and the reſt, at a certain point of its deſcent, is thrown out again by the cen-trifugal force. This appears to have been the caſe in the three firſt experiments of the foregoing ſpecimen ; but as the velocity, when this begins to happen, depends upon the form of the buckets, as well as other circumſtances, *the utmost velocity of overshot wheels is not to be determined generally :* and, indeed, it is the leſs neceſſary in prac-tice, as it is in this circumſtance incapable of producing any *mechanical effect,* for reaſons already given.

VI. Concerning the greateſt Load that an Overſhot Wheel can overcome.

The greatest load an overshot wheel will overcome, considered abstractedly, is unlimited or infinite : for as the buckets may be of any given capacity, the more the

wheel

wheel is loaded, the flower it turns; but the flower it turns, the more will the buckets be filled with water; and confequently though the diameter of the wheel and quantity of water expended, are both limited, yet no refiftance can be affigned, which it is not able to overcome: but in practice we always meet with fomething that prevents our getting into infinitefimals; for when we really go to work to build a wheel, the buckets muft neceffarily be of fome given capacity; and confequently *such a resistance will stop the wheel, as is equal to the effort of all the buckets in one semi-circumference filled with water.*

The ftructure of the buckets being given; the quantity of this effort may be affigned; but is not of much confequence to the practice, as in this cafe alfo the wheel lofes its power; for though here be the exertion of gravity upon a given quantity of water, yet being prevented by a counterbalance from moving, is capable of producing no *mechanical effect*, according to our definition. But, in reality, an overfhot wheel generally ceafes to be ufeful before it is loaded to that pitch; for *when it meets with such a resistance as to diminish its velocity to a certain degree, its motion becomes irregular; yet this never happens until the velocity of the circumference is less than 2 feet per second, where the resistance is equable,* as appears not only from the preceding fpecimen, but from experiments on larger wheels.

Scholium.

Having now examined the different effects of the power of water, when acting by its *impulse*, and by its *weight*, under the titles of *undershot* and *overshot* wheels; we might naturally proceed to examine the effects when the impulfe and weight are combined, as in the feveral kinds of *breast wheels, &c.* but, what has been already delivered being carefully attended to, the application of the fame principles in thefe mixt cafes will be eafy, and reduce what I have to fay on this head into a narrow compafs: for all kinds of wheels where the water cannot defcend through a given fpace, unlefs the wheel moves therewith, are to be confidered of the nature of an overfhot wheel, according to the perpendicular height that the water defcends from; and all thofe that receive the impulfe or fhock of the water, whether in an horizontal, perpendicular, or oblique direction, are to be confidered as underfhots. And therefore a wheel which the water ftrikes at a certain point below the furface of the head, and after that defcends in the arch of a circle, preffing by its gravity upon the wheel; *the effect of such a wheel will be equal to the effect of an undershot, whose head is equal to the difference of level between*

the

the surface of the water in the reservoir and the point where it strikes the wheel, added to that of an overshot, whose height is equal to the difference of level, between the point where it strikes the wheel and the level of the tail-water. It is here fuppofed, that the wheel receives the fhock of the water at right angles to its radii; and that the velocity of its circumference is properly adapted to receive the utmoft advantage of both thefe powers; otherwife a reduction muft be made on that account.

Many obvious and confiderable improvements upon the common practice naturally offer themfelves, from a due confideration of the principles here eftablifhed, as well as many popular errors fhew themfelves in view: but as my prefent purpofe extends no farther than the laying down fuch general rules as will be found to anfwer in practice, I leave the particular application to the intelligent artift, and to the curious in thefe matters.

PART

Plate 6. page 11.

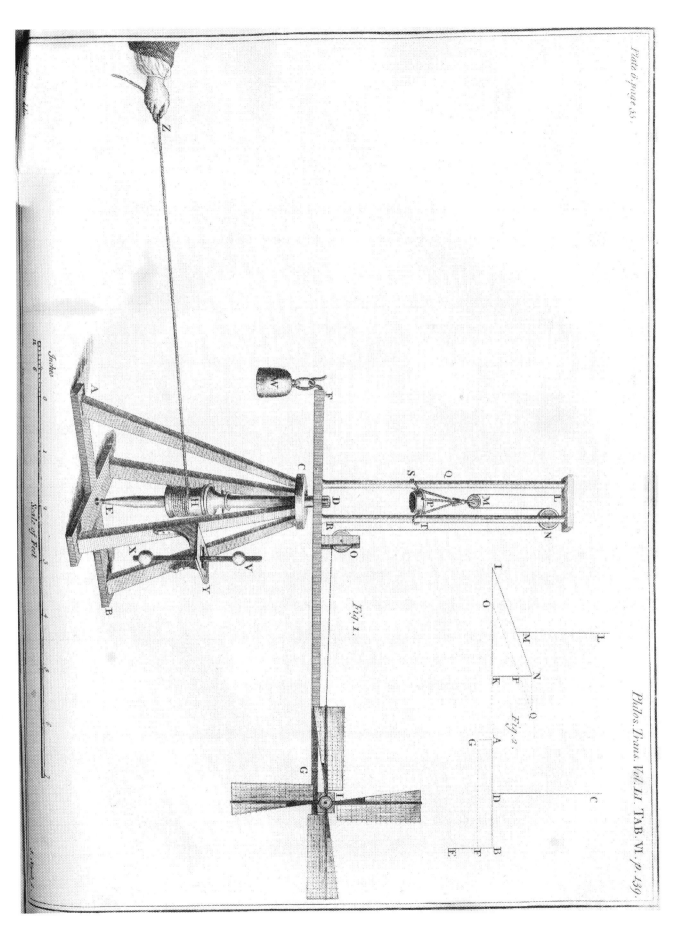

Fig.1.

Fig.2.

Inches

Scale of Feet

PART III.

On the Construction and Effects of Windmill Sails.

Read before the Royal Society, 31ft May and 14th June, 1759.

IN trying experiments on wind mill fails, the wind itfelf is too uncertain to anfwer the purpofe; we muft therefore have recourfe to an artificial wind.

This may be done two ways; either by caufing the air to move againft the machine, or the machine to move againft the air. To caufe the air to move againft the machine, in a fufficient volume, with fteadinefs and the requifite velocity, is not eafily put in practice: to carry the machine forward in a right line againft the air, would require a larger room than I could conveniently meet with. What I found moft practicable, therefore, was to carry the axis, whereon the fails were to be fixed, progreffively round in the circumference of a large circle. Upon this idea * a machine was con‑ ftructed, as follows:

PLATE VI. Fig. 1.

A B C is a pyramidical frame for fupporting the moving parts.

D E is an upright axis, whereon is framed

F G, an arm for carrying the fails at a proper diftance from the centre of the upright axis.

---

* Some years ago Mr. Roufe, an ingenious gentleman of Harborough, in Leicestershire, set about trying experiments on the velocity of the wind, and force thereof upon plain surfaces and wind mill sails: and, much about the same time, Mr. Ellicott contrived a machine for the use of the late celebrated Mr. B. Robins, for trying the resistance of plain surfaces moving through the air. The machines of both these gentlemen were much alike, though at that time totally unacquainted with each other's inquiries. But it often happens, that when two persons think justly upon the same subject, their ex‑ periments are alike. This machine was also built upon the same idea as the foregoing; but differed in having the hand for the first mover, with a pendulum for its regulator, instead of a weight, as in the former; which was certainly best for the purposes of measuring the impulse of the wind, or resistance of plains: but the latter is more applicable to experiments on windmill sails; because every change of position of the same sails will occasion their meeting the air with a different velocity, though urged by the same weight.

H is

H is a barrel upon the upright axis, whereon is wound a cord; which, being drawn by the hand, gives a circular motion to the axis, and to the arm F G; and thereby carries the axis of the fails in the circumference of a circle, whofe radius is D I, caufing thereby the fails to ftrike the air, and turn round upon their own axis.

At L is fixed the end of a fmall line, which paffing through the pullies M N O, terminates upon a fmall cylinder or barrel upon the axis of the fails, and, by winding thereon raifes

P the fcale, wherein the weights are placed for trying the power of the fails.   This fcale, moving up and down in the direction of the upright axis, receives no difturbance from the circular motion.

Q R two parallel pillars ftanding upon the arm F G, for the purpofe of fupporting and keeping fteady the fcale P; which is kept from fwinging by means of

S T, two fmall chains, which hang loofely round the two pillars.

W is a weight for bringing the centre of gravity of the moveable part of the machine into the centre of motion of the axis D E.

V X is a pendulum, compofed of two balls of lead, which are moveable upon a wooden rod, and thereby can be fo adjufted, as to vibrate in any time required.   This pendulum hangs upon a cylindrical wire, whereon it vibrates, as on a rolling axis.

Y is a perforated table for fupporting the axis of the pendulum.

Note, The pendulum being fo adjufted, as to make two vibrations in the time that the arm F G is intended to make one turn; the pendulum being fet a vibrating, the experimenter pulls by the cord Z, with fufficient force to make each half revolution of the arm to correfpond with each vibration, as equal as poffible during the number of vibrations that the experiment is intended to be continued.   A little practice renders it eafy to give motion thereto with all the regularity that is neceffary.

Specimen

Specimen of a Set of Experiments.

Radius of the fails   -   -   -   -   -   21 inches.
Length of ditto in the cloth -   -   -   -   18
Breadth of ditto -   -   -   -   -   -   5,6
* { Angle at the extremity   -   -   -   10 degrees.
   { Ditto at the greateft inclination -   -   -   25
20 turns of the fails raifed the weight   -   -   11,3
Velocity of the centrè of the fails, in the cir-
   cumference of the great circle, in a fe- } 6 feet 0 inches.
   cond   -   -   -   -   -   - }
Continuance of the experiment   -   -   -   52 feconds.

No.	Weight in the fcale.	Turns.	Product.
1	0 *lb.*	108	0
2	6	85	510
3	$6\frac{1}{2}$	81	$526\frac{1}{2}$
4	7	78	546
5	$7\frac{1}{2}$	73	$547\frac{1}{2}$ maximum.
6	8	65	520
7	9	0	0

N. B. The weight of the fcale and pulley was 3 oz.; and that 1 oz. fufpended upon one of the radii, at $1\frac{1}{2}$ inches from the centre of the axis, juft overcame the friction, fcale and load of $7\frac{1}{2}$ lb.; and placed at $14\frac{12}{18}$ inches, overcame the fame refiftances with 9 lb. in the fcale.

Reduction of the preceding Specimen.

No. 5, being taken for the maximum, the weight in the fcale was 7 lb. 8 oz. which, with the weight of the fcale and pulley 3 oz. makes 7 lb. 11 oz. equal to 123 oz.;

---

* In all the following experiments the angle of the sails is accounted from the plane of their motion; that is, when they stand at right angles to the axis, their angle is denoted 0°, this notation being agreeable to the language of practitioners, who call the angle so denoted, the weather of the sail; which they denominate greater or less, according to the quantity of this angle.

I

this,

this, added to the friction of the machinery, the sum is the whole resistance *. The friction of the machinery is thus deduced: since 20 turns of the sails raised the weight 11,3 inches, with a double line, the radius of the cylinder will be .18 of an inch; but had the weight been raised by a single line, the radius of the cylinder being half the former, viz. .09, the resistance would have been the same: we shall therefore have this analogy; as half the radius of the cylinder, is to the length of the arm where the small weight was applied; so is the weight applied to the arm, to a fourth weight, which is equivalent to the sum of the whole resistance together; that is, .09 : 12,5 :: 1 oz. : 139 oz. this exceeds 123 oz. the weight in the scale, by 16 oz. or 1 lb. which is equivalent to the friction; and which, added to the above weight of 7 lb. 11 oz. makes 8 lb. 11 oz. = 8,69 lb. for the sum of the whole resistance; and this, multiplied by 73 turns, makes a product of 634, which may be called the representative of the *effect* produced.

In like manner, if the weight 9 lb. which caused the sails to rest after being in motion, be augmented by the weight of the scale and its relative friction, it will become 10,37 lb. The result of this specimen is set down in No. 12, of Table III. and the result of every other set of experiments therein contained were made and reduced in the same manner.

---

* The resistance of the air is not taken into the account of resistance, because it is inseparable from the application of the power.

TABLE.

TABLE III.

Containing **Nineteen** Sets of Experiments on Windmill Sails of various Structures, Positions, and Quantities of Surfaces.

The kind of sails made use of.	Number.	Angle at the extremities.	Greatest angle.	Turns of the sails unloaded.	Turns of ditto at the maximum.	Load at the maximum.	Greatest load.	Product.	Quantity of surface.	Ratio of greatest velocity to the velocity at maximum.	Ratio of greatest load to the load at maximum.	Ratio of surface to the product.
		°	°			lb.	lb.		Sq. In.			
Plain sails at an angle of 55°	1	35	35	66	42	7,56	12,59	318	404	10:7	10:6	10:7,9
Plain sails weathered according to the common practice.	2	12	12		70	6,3	7,56	441	404		10:8,3	10:10,1
	3	15	15	105	69	6,72	8,12	464	404	10:6,6	10:8,3	10:10,15
	4	18	18	96	66	7,0	9,81	462	404	10:7,	10:7,1	10:10,15
Weathered according to *Maclaurin's* theorem.	5	9	26½		66	7,0			462	404			10:11,4
	6	12	29½		70¹	7,35		518	404			10:12,8
	7	15	32½		63⅓	8,3		527	404			10:13,
Sails weathered in the *Dutch* manner, tried in various positions.	8	0	15	120	93	4,75	5,31	442	404	10:7,7	10:8,9	10:11,
	9	3	18	120	79	7,0	8,12	553	404	10:6,6	10:8,6	10:13,7
	10	5	20		78	7,5	8,12	585	404		10:9,2	10:14,5
	11	7½	22½	113	77	8,3	9,81	639	404	10:6,8	10:8,5	10:15,8
	12	10	25	108	73	8,69	10,37	634	404	10:6,8	10:8,4	10:15,7
	13	12	27	100	66	8,41	10,94	580	404	10:6,6	10:7,7	10:14,4
Sails weathered in the *Dutch* manner, but *enlarged* towards the extremities.	14	7½	22½	123	75	10,65	12,59	799	505	10:6,1	10:8,5	10:15.8
	15	10	25	117	74	11,08	13,69	820	505	10:6,3	10:8,1	10:16,2
	16	12	27	114	66	12,09	14,23	799	505	10:5,8	10:8,4	10:15,8
	17	15	30	96	63	12,09	14,78	762	505	10:6,6	10:8,2	10:15,1
Eight sails being *sectors of ellipses* in their best positions.	18	12	22	105	64½	16,42	27,87	1059	854	10:6,1	10:5,9	10:12,4
	19	12	22	99	64½	18,06		1165	1146	10:5,0		10:10,1
	1.	2.	3.	4.	5.	6.	7.	8.	9.	10.	11.	12.

Observations and Deductions from the preceding Experiments.

I. Concerning the beſt Form and Poſition of Windmill-Sails.

In Table III. N°. 1. is contained the reſult of a ſet of experiments upon ſails ſet at the angle which the celebrated Monſ. Parint, and ſucceeding geometricians for many years, held to be the beſt; *viz.* thoſe whoſe planes make an angle 55° nearly with the axis; the complement whereof, or angle that the plane of the ſail makes with the plane of their motion, will therefore be 35°, as ſet down in columns 2. and 3. Now if we multiply their number of turns by the weight they lifted, when working to the greateſt advantage, as ſet down in columns 5. and 6. and compare this product (col. 8.) with the other products contained in the ſame column, inſtead of being the greateſt, it turns out the leaſt of all the reſt. But if we ſet the angle of the ſame planes at ſomewhat leſs than half the former, or at any angle from 15° to 18°, as in N°. 3. and 4. that is, from 72° to 75° with the axis, the product will be increaſed in the ratio of 31 : 45; and this is the angle moſt commonly made uſe of by practioners, when the ſurfaces of the ſails are planes.

If nothing more was intended than to determine the moſt efficacious angle to make a mill acquire motion from a ſtate of reſt, or to prevent it from paſſing into reſt from a ſtate of motion, we ſhall find the poſition of N°. 1. the beſt; for if we conſult col. 7. which contains the leaſt weights, that would make the ſails paſs from motion to reſt, we ſhall find that of N°. 1. (relative to the quantity of cloth) the greateſt of all. But if the ſails are intended, with given dimenſions, to produce the greateſt effect poſſible in a given time, we muſt entirely reject thoſe of N°. 1. and, *if we are confined to the uſe of planes, conform ourſelves to ſome angle between* N°. 3. *and* 4. *that is, not leſs than* 72°, *or greater than* 75 , *with the axis.*

The late celebrated Mr. Maclaurin has judiciouſly diſtinguiſhed between the action of the wind upon a ſail at reſt, and a ſail in motion; and, in conſequence, as the motion is more rapid near the extremities than towards the centre, that the angle of the different parts of the ſail, as they recede from the centre, ſhould be varied. For this purpoſe he has furniſhed us with the following theorem *. " Suppoſe the velocity of the wind to " be repreſented by a, and the velocity of any given part of the ſail to be denoted by c;

---

* Maclaurin's account of Sir Iſaac Newton's philoſophical diſcoveries, p. 176, art. 29.

" then

" then the effort of the wind upon that part of the fail will be greateſt when the tangent
" of the angle, in which the wind ſtrikes it, is to radius as

$$" \sqrt{2 + \frac{9cc}{4aa} + \frac{3c}{2a}} \text{ to } 1."$$

This theorem then aſſigns the law, by which the angle
is to be varied according to the velocity of each part of the fail to the wind: but as it is
left undetermined what velocity any one given part of the fail ought to have in reſpect
to the wind, the angle that any one part of the fail ought to have, is left undetermined
alſo; ſo that we are ſtill at a loſs for the proper *data* to apply the theorem. However,
being willing to avail myſelf thereof, and conſidering that any angle from 15° to 18°
was beſt ſuited to a plane, and of conſequence to the beſt mean angle, I made the fail,
at the middle diſtance between the centre and the extremity, to ſtand at an angle of
15° 41 with the plane of the motion: in which caſe the velocity of that part of the
fail, when loaded to a *maximum*, would be equal to that of the wind, or $c=a$. This
being determined, the reſt were inclined according to the theorem, as follows:

			Angle with the axis.		Angle of weather.	
Parts of the radius from the centre.	$\frac{1}{6}$	$c=\frac{1}{3}a$	63° 26′		26° 34′	
	$\frac{1}{3}$	$c=\frac{2}{3}a$	69 54		20 6	
	$\frac{1}{2}$	$c=a$	74 19		15 41	middle.
	$\frac{2}{3}$	$c=1\frac{1}{3}a$	77 20		12 40	
	$\frac{5}{6}$	$c=1\frac{1}{2}a$	79 27		10 33	
	1	$c=2a$	81 0		9 0	extremity.

The reſult hereof was according to N°. 5, being nearly the ſame as the plane fails,
in their beſt poſition: but being turned round in their ſockets, ſo that every part of
each fail ſtood at an angle of 3°, and afterwards of 6°, greater than before, that is,
their extremities being moved from 9° to 12° and 15°, the products were advanced
to 518 and 527 reſpectively. Now from the ſmall difference between thoſe two products, we may conclude, that they were nearly in their beſt poſition, according to
N°. 7, or ſome angle between that and N°. 6: but from theſe as well as the plane
fails and others, we may alſo conclude, that *a variation in the angle of a degree or
two makes very little difference in the effect, when the angle is near upon the
beſt.*

It is to be obſerved, that a fail inclined by the preceding rule will expoſe a convex
ſurface to the wind: whereas the Dutch, and all our modern mill-builders, though
they

make the angle to diminish, in receding from the centre towards the extremity, yet constantly do it in such manner, as that the surface of the sail may be concave towards the wind. In this manner the sails made use of in N°. 8, 9, 10, 11, 12, and 13, were constructed; the middle of the sail making an angle with the extreme bar of 12°; and the greatest angle (which was about ⅐ of the radius from the centre) of 15° therewith. Those sails being tried in various positions, the best appears to be that of N°. 11, where the extremities stood at an angle of 7°½ with the plane of motion, the product being 639: greater than that of those made by the theorem in the ratio of 9 : 11, and double to that of N°. 1; and this was the greatest product that could be procured without an augmentation of surface. Hence it appears, that *when the wind falls upon a concave surface, it is an advantage to the power of the whole, though every part, taken separately, should not be disposed to the best advantage* *

Having thus obtained the best position of the sails, or manner of weathering, as it is called by the workmen, the next point was to try what advantage could be made by an addition of surface upon the same radius. For this purpose, the sails made use of had the same weather as those N°. 8 to 13, with an addition to the leading side of each of a triangular cloth, whose height was equal to the height of the sail, and whose base was equal to half the breadth: of consequence the increase of surface upon the whole was one-fourth part, or as 4 : 5. Those sails, by being turned round in their sockets, were tried in four different positions, specified in N°. 14, 15, 16, and 17; from whence it appears, that the best was when every part of the sail made a greater angle by 2°½, with the plane of the motion, than those without the addition, as appears by N° 15, the product being 820: this exceeds 639 more than in the ratio of 4 : 5, or that of the increase of cloth. Hence it appears, that *a broader sail requires a*

---

* By several trials in large I have found the following angles to answer as well as any. The radius is supposed to be divided into six parts and one-sixth, reckoning from the centre, is called one, the extremity being denoted six.

N°.			Angle with the axis.				Angle with the plane of the motion.
1	•	-	72°	-	•	-	18°
2	-	-	71	-	•	-	19
3	-	•	72	-	•	-	18 middle.
4	•	-	74	-	•	-	16
5	-	-	77½	-	•	-	12½
6	-	-	83	-	•	•	7 extremity.

greater

greater angle ; and *that when the sail is broader at the extremity, than near the centre, this shape is more advantageous than that of a parallelogram* *

Many have imagined, that the more fail, the greater the advantage, and have therefore propofed to fill up the whole area: and by making each fail a fector of an ellipfis, according to Monfieur Parint, to intercept the whole cylinder of wind, and thereby to produce the greateft effect poffible.

We have therefore proceeded to inquire how far the effect could be increafed by a further enlargement of the furface, upon the fame radius of which N°. 18 and 19 are fpecimens. The furfaces indeed were not made planes, and fet at an angle of 35°, as Parint propofed ; becaufe, from N°. 1, we learn, that this pofition has nothing to do, when we intend them to work to the greateft advantage. We therefore gave them fuch an angle as the preceding experiments indicated for fuch fort of fails, viz. 12° at the extremity, and 22 for the greateft weather. By N°. 18 we have the product 1059, greater than N°. 15, in the ratio of 7 : 9 ; but then the augmentation of cloth is almoft 7 : 12. By N°. 19, we have the product 1165, that is greater than N°. 15, as 7 : 10 ; but the augmentation of cloth is nearly as 7 : 16 ; confequently had the fame quantity of cloth as in N°. 18, been difpofed in a figure fimilar to that of N°. 15, inftead of the product 1059, we fhould have had the product 1336 ; and in N°. 19, inftead of the product 1165, we fhould have had a product of 1860 ; as will be further made appear in the courfe of the following deductions. Hence it appears, that beyond a certain degree, the more the area is crouded with fail, the lefs effect is produced in proportion to the furface : and, by purfuing the experiments ftill further, I found, that though in N°. 19, the furface of all the fails together were not more than feven-eighths of the circular area containing them, yet a further addition rather diminifhed than increafed the effect. *So that when the whole cylinder of wind is intercepted, it does not then produce the greatest effect for want of proper interstices to escape.*

---

* The figure and proportion of the enlarged fails, which I have found beft to anfwer in large, are reprefented in the figure, Plate VI. where the extreme bar is one-third of the radius (or whip, as it is called by the workmen), and is divided by the whip in the proportion of 3 to 5. The triangular or leading fail is covered with board from the point downwards one-third of its height, the reft with cloth as ufual. The angles of weather in the preceding note are beft for the enlarged fails alfo ; for in practice it is found, that the fails had better have too little than much weather.

It is certainly defirable that the fails of windmills fhould be as fhort as poffible; but at the fame time it is equally defirable, the quantity of cloth fhould be the leaft that may be, to avoid damage by fudden fqualls of wind. The beft ftructure, therefore, for large mills, is that where the quantity of cloth is the greateft, in a given circle, that can be: on this condition, that the effect holds out in proportion to the quantity of cloth; for otherwife the effect can be augmented in a given degree by a leffer increafe of cloth upon a larger radius, than would be required, if the cloth was increafed upon the fame radius. The moft ufeful figure therefore for practice, is that of N°. 9 or 10, as has been experienced upon feveral mills in large.

TABLE IV.

Containing the Refult of fix Sets of Experiments, made for determining the Difference of Effect, according to the different Velocity of the Wind.

N. B. The fails were of the fame fize and kind as thofe of N°. 10, 11, and 12, Table III.

Continuance of the Experiment one Minute.

Number.	Angle at the extremity.	Velocity of the wind in a fecond.	Turns of the fails unloaded.	Turns of the fails at maximum.	Load at the maximum.	Greatest load.	Product.	Maximum load for the half velocity.	Turns of the fails therewith.	Product of leffer load and greater velocity.	Ratio of the two products.	Ratio of the greateft velocity to the velocity at a maximum.	Ratio of the greateft load to the load at a maximum.
		f. in.			lb.	lb.							
1	5°	4 4½	96	66	4,47	5,37	295	—	—	—	— —	10:6,9	10:8,3
2	5	8 9	207	122	16,42	18,06	2003	4,47	180	805	10 : 27,3	10:5,9	10:9,1
3	7½	4 4½	—	65	4,62	—	300	—	—	—	— —	—	—
4	7½	8 9	—	130	17,52	—	2278	4,62	180	832	10 : 27,8	—	—
5	10	4 4½	91	61	5,03	5,87	307	—	—	—	— —	10:6,7	10:8,5
6	10	8 9	178	110	18,61	21,34	2047	5,03	158	795	10 : 26	10:6,2	10:8,7
1	2	3	4	5	6	7	8	9	10	11	12	13	14

II. *Concerning*

II. Concerning the Ratio between the Velocity of Windmill Sails unloaded, and their Velocity when loaded to a *Maximum*.

Thofe ratios, as they turned out in experiments upon different kinds of fails, and with different inclinations (the velocity of the wind being the fame) are contained in column 10 of Table III. where the extremes differ from the ratio of 10 : 7,7 to that of 10 : 5,8 ; but *the most general ratio of the whole will be nearly as* 3 : 2. This ratio alfo agrees fufficiently near with experiments where the velocity of the wind was different, as in thofe contained in Table IV. column 13, in which the ratios differ from 10   6,9 to that of 10 : 5,9. However, it appears in general, that where the power is greater, whether by an enlargement of furface, or a greater velocity of the wind, that the fecond term of the ratio is lefs.

III. Concerning the Ratio between the greateft Load that the Sails will bear without ftopping, or what is nearly the fame Thing, between the leaft Load that will ftop the Sails, and the Load at the *Maximum*.

Thofe ratios for different kinds of fails and inclinations, are collected in column 11, Table III. where the extremes differ from the ratio of 10   6 to that of 10 : 9,2 ; but taking in thofe fets of experiments only, where the fails refpectively anfwered beft, *the ratio's will be confined between that of* 10 : 8 *and of* 10 : 9 ; *and at a medium about* 10 : 8,3 *or of* 6 : 5. This ratio alfo agrees nearly with thofe in column 14 of Table IV. However it appears, upon the whole, that in thofe inftances, where the angle of the fails or quantity of cloth were greateft, that the fecond term of the ratio was lefs.

IV. Concerning the Effects of Sails, according to the different Velocity of the Wind.

Maxim I. The velocity of windmill fails, whether unloaded, or loaded, fo as to produce a *maximum*, is nearly as the velocity of the wind, their fhape and pofition being the fame.

This appears by comparing together the refpective numbers of columns 4 and 5, Table IV. wherein thofe of numbers 2, 4, and 6, ought to be double of numbers 1, 3, and 5 : but as the deviation is no where greater than what may be imputed to

K   the

the inaccuracy of the experiments themfelves, and hold good exactly in numbers 3 and 4; which fets were deduced from the medium of a number of experiments, carefully repeated the fame day, and on that account are moſt to be depended upon, we therefore conclude the maxim true.

Maxim II. The load at the *maximum* is nearly, but fomewhat leſs than, as the fquare of the velocity of the wind, the ſhape and poſition of the fails being the fame.

This appears by comparing together the numbers in column 6, Table IV. wherein thoſe of numbers 2, 4, and 6, (as the velocity is double), ought to be quadruple of thoſe of numbers 1, 3, and 5; inſtead of which they fall ſhort, number 2 by $\frac{1}{14}$, number 4 by $\frac{1}{15}$, and number 6 by $\frac{1}{13}$ part of the whole. The greateſt of thoſe deviations is not more conſiderable than might be imputed to the unavoidable errors in making the experiments : but as thoſe experiments, as well as thoſe of the greateſt load, all deviate the fame way; and alſo coincide with fome experiments communicated to me by Mr. Rouſe upon the reſiſtance of planes; I am led to ſuppoſe a fmall deviation, whereby the load falls ſhort of the fquares of the velocity; and fince the experiments No. 3 and 4, are moſt to be depended upon, we muſt conclude, that when the velocity is double, the load falls ſhort of its due proportion by $\frac{1}{15}$, or, for the fake of a round number, by about $\frac{1}{16}$ part of the whole.

Maxim III. The effects of the fame fails at a *maximum* are nearly, but fomewhat leſs than, as the cubes of the velocity of the wind.

It has already been proved, Maxim 1ſt, that the velocity of fails at the *maximum*, is nearly as the velocity of the wind; and by Maxim 2d, that the load at the *maximum* is nearly as the fquare of the fame velocity : if thoſe two *maximums* would hold preciſely, it would be a confequence that the effect would be in a triplicate ratio thereof : how this agrees with experiment will appear by comparing together the products in column 8, of Table IV. wherein thoſe of No. 2, 4, and 6, (the velocity of the wind being double) ought to be octuple of thoſe of No. 1, 3, and 5, inſtead of which they fall ſhort, No. 2 by $\frac{1}{7}$, No. 4 by $\frac{1}{20}$, and No. 6, by $\frac{1}{6}$ part of the whole. Now, if we rely on No. 3 and 4, as the turns of the fails are as the velocity of the wind; and fince the load of the *maximum* falls ſhort of the fquare of the velocity by about $\frac{1}{16}$ part of the whole : the product made by the multiplication of the turns into the load, muſt alſo fall ſhort of the triplicate ratio by about $\frac{1}{16}$ part of the whole product.

Maxim

Maxim IV. The load of the fame fails at the *maximum* is nearly as the fquares, and their effect as the cubes of their number of turns in a given time.

This maxim may be efteemed a confequence of the three preceding; for if the turns of the fails are as the velocity of the wind, whatever quantities are in any given ratio of the velocity of the wind, will be in the fame given ratio of the turns of the fails: and therefore, if the load at the *maximum* is as the fquare, or the effect as the cube, of the velocity of the wind, wanting $\frac{1}{10}$ part when the velocity is double; the load at the *maximum* will alfo be as the fquare, and the effect as the cube, of the number of turns of the fails in a given time, wanting in like manner $\frac{1}{10}$ part when the number of turns are double in the fame time. In the prefent cafe, if we compare the loads at the *maximum*, column 6, with the fquares of the number of turns, column 5, of No. 1 and 2, 5 and 6, or the products of the fame numbers column 8, with the cubes of the number of turns column 5, inftead of falling fhort, as No. 3 and 4, they exceed thofe ratios: but as the fets of experiments No. 1 and 2, of 5 and 6, are not to be efteemed of equal authority with thofe of No. 3 and 4, we muft not rely upon them further than to obferve, that *in comparing the gross effects of large machines, the direct proportion of the squares and cubes respectively, will hold as near as the effects themselves can be observed;* and therefore be fufficient for practical eftimation, without any allowance.

Maxim V. When fails are loaded fo as to produce a *maximum* at a given velocity, and the velocity of the wind increafes, the load continuing the fame; 1ft, The increafe of effect, when the increafe of the velocity of the wind is fmall, will be nearly as the fquares of thofe velocities: 2dly, When the velocity of the wind is double, the effects will be nearly as 10: $27\frac{1}{2}$: But, 3dly, When the velocities compared, are more than double of that where the given load produces a *maximum*, the effects increafe nearly in a fimple ratio of the velocity of the wind.

It has already been proved, maxim 1ft and 2d, that when the velocity of the wind is increafed, the turns of the fails will increafe in the fame proportion, even when oppofed by a load as the fquare of the velocity: and therefore if wanting the oppofition of an increafe of load, as the fquare of the velocity, the turns of the fails will again be increafed in a fimple ratio of the velocity of the wind on that account alfo; that is, the load continuing the fame, the turns of the fails in a given time will be as the fquare of the velocity of the wind; and the effect, being in this cafe as the turns of

the

the fails will be as the fquare of the velocity of the wind alfo; but this muft be underftood only of the firft increments of the velocity of the wind: for,

2dly, As the fails will never acquire above a given velocity in relation to the wind, though
the load were diminifhed to nothing; when the load continues the fame, the more the velocity of the wind increafes (though the effect will continue to increafe) yet the more it
will fall fhort of the fquare of the velocity of the wind; fo that when the velocity of
the wind is double, the increafe of effect, inftead of being as 1 : 4, according to the
fquares, it turns out as 10 : 27½, as thus appears. In Table IV. column 9, the loads
of No. 2, 4, and 6, are the fame as the *maximum* loads in column 6, of No. 1, 3,
and 5. The number of turns of the fails with thofe loads, when the velocity of the
wind is double, are fet down in column 10, and the products of their multiplication
in column 11: thofe being compared with the products of No. 1, 3, and 5, column 8,
furnifh the ratios fet down in column 12, which at a medium (due regard being had
to No. 3 and 4.) will be nearly as 10 : 27½. 3dly. The load continuing the fame,
grows more and more inconfiderable, refpecting the power of the wind as it increafes
in velocity; fo that the turns of the fails grow nearer and nearer a coincidence with
their turns unloaded; that is, nearer and nearer to the fimple ratio of the velocity of
the wind. When the velocity of the wind is double, the turns of the fails, when
loaded to a *maximum*, will be double alfo; but, *unloaded*, will be no more than
triple, by deduction 2d: and therefore the product could not have increafed beyond
the ratio of 10 : 30 (inftead of 10 : 27½) even fuppofing the fails not to have been
retarded at all by carrying the *maximum* load for the half velocity. Hence we fee,
that when the velocity of the wind exceeds the double of that, where a conftant load
produces a *maximum*, that the increafe of effect, which follows the increafe of the
velocity of the fails, will be nearly as the velocity of the wind, and ultimately in that
ratio precifely. Hence alfo we fee that windmills, fuch as the different fpecies for
raifing water for drainage, &c. lofe much of their full effect, when acting againft one
invariable oppofition.

V. Concerning the Effects of Sails of different Magnitudes the Structure and Pofition
being fimilar, and the Velocity of the Wind the fame.

Maxim 6. In fails of a fimilar figure and pofition, the number of turns in a given
time will be reciprocally as the radius or length of the fail.

The

The extreme bar having the fame inclination to the plane of its motion, and to the wind; its velocity at a *maximum* will always be in a given ratio to the velocity of the wind; and therefore, whatever be the radius, the abfolute velocity of the extremity of the fail will be the fame: and this will hold good refpecting any other bar, whofe inclination is the fame, at a proportionable diftance from the centre; it therefore follows, that the extremity of all fimilar fails, with the fame wind, will have the fame abfolute velocity; and therefore take a fpace of time to perform one revolution in proportion to the radius; or, which is the fame thing, the number of revolutions in the fame given time, will be reciprocally as the length of the fail.

Maxim 7. The load at a *maximum* that fails of a fimilar figure and pofition will overcome, at a given diftance from the centre of motion, will be as the cube of the radius.

Geometry informs us, that in fimilar figures the furfaces are as the fquares of their fimilar fides; of confequence the quantity of cloth will be as the fquare of the radius: alfo in fimilar figures and pofitions, the impulfe of the wind, upon every fimilar fection of the cloth, will be in proportion to the furface of that fection; and confequently, the impulfe of the wind upon the whole, will be as the furface of the whole: but as the diftance of every fimilar fection, from the centre of motion, will be as the radius; the diftance of the centre of power of the whole, from the centre of motion, will be as the radius alfo; that is, the lever by which the power acts, will be as the radius: as therefore the impulfe of the wind, refpecting the quantity of cloth, is as the fquare of the radius, and the lever, by which it acts, as the radius fimply; it follows, that the load which the fails will overcome, at a given diftance from the centre, will be as the cube of the radius.

Maxim 8. The effect of fails of fimilar figure and pofition, are as the fquare of the radius.

By maxim 6. it is proved, that the number of revolutions made in a given time, are as the radius inverfely. Under maxim 7. it appears, that the length of the lever, by which the power acts, is as the radius directly; therefore thefe equal and oppofite ratios deftroy one another: but as in fimilar figures the quantity of cloth is as the fquare of the radius, and the action of the wind is in proportion to the quantity of cloth, as alfo appears under maxim 7; it follows that the effect is as the fquare of the radius.

COROL.

COROL. 1. Hence it follows, that augmenting the length of the sail, without augmenting the quantity of cloth, does not increase the power; because what is gained by the length of the lever, is lost by the slowness of the rotation.

COROL. 2. If sails are increased in length, the breadth remaining the same, the effect will be as the radius.

VI. Concerning the Velocity of the Extremities of Windmill Sails, in respect to the Velocity of the Wind.

Maxim 9. The velocity of the extremities of Dutch sails, as well as of the enlarged sails, in all their usual positions when unloaded, or even loaded to a *maximum*, are considerably quicker than the velocity of the wind.

The Dutch sails unloaded, as in Table III. No. 8. made 120 revolutions in 52″: the diameter of the sails being 3 feet 6 inches, the velocity of their extremities will be 25,4 feet in a second; but the velocity of the wind producing it, being 6 feet in the same time, we shall have 6 : 25,4 : : 1 : 4,2; in this case therefore, the velocity of their extremities was 4,2 times greater than that of the wind. In like manner, the relative velocity of the wind, to the extremities of the same sails, when loaded to a *maximum*, making then 93 turns in 52″, will be found to be as 1 : 3,3; or 3,3 times quicker than that of the wind.

The following table contains 6 examples of Dutch sails, and 4 examples of the enlarged sails, in different positions, but with the constant velocity of the wind of 6 feet in a second, from Table III.: and also 6 examples of Dutch sails in different positions, with different velocities of the wind from Table IV.

TABLE

TABLE V.

Containing the Ratio of the Velocity of the Extremities of Windmill Sails to the Velocity of the Wind.

Number.	No. of Table III. and IV.	Angle at the Extremity.	Velocity of the Wind in a Second.		Ratio of the Velocity of the Wind, and Extremities of the Sails.		
					unloaded.	loaded.	
1	8	0°	6^f	.0^{in}	1 : 4,2	1 : 3,3	
2	9	3	6	0	1 : 4,2	1 : 2,8	
3	10	5	6	0	—— ·——	1 : 2,75	From Table III.
4	11	7	6	0	1 : 4,	1 : 2,7	
5	12	10	6	0	1 : 3,8	1 : 2,6	
6	13	12	6	0	1 : 3,5	1 : 2,3	
7	14	7$\frac{1}{2}$	6	0	1 : 4,3	1 : 2,6	
8	15	10	6	0	1 : 4,1	1 : 2,6	
9	16	12	6	0	1 : 4,	1 : 2,3	
10	17	15	6	0	1 : 3,35	1 : 2,2	
11	1	5	4	4$\frac{1}{2}$	1 : 4,	1 : 2,8	
12	2	5	8	9	1 : 4,3	1 : 2,6	
13	3	7$\frac{1}{2}$	4	4$\frac{1}{2}$	—— ——	1 : 2,8	From Tab. IV.
14	4	7$\frac{1}{2}$	8	9	—— ——	1 : 2,7	
15	5	10	4	4$\frac{1}{2}$	1 : 3,8	1 : 2,6	
16	6	10	8	9	1 : 3,4	1 : 23	
1	2	3	4		5	6	

It appears from the preceding collection of examples, that when the extremities of the Dutch sails are parallel to the plane of motion, or at right angles to the wind, and to the axis, as they are made according to the common practice in England, that their velocity, unloaded, is above 4 times, and loaded to a *maximum*, above 3 times greater than that of the wind: but that when the Dutch sails, or enlarged sails, are in their best positions, their velocity unloaded is 4 times, and loaded to a *maximum*, at a medium the

Dutch

Dutch fails are 2,7, and the enlarged fails 2,6 times greater than the velocity of the wind. Hence we are furnished with a method of knowing the velocity of the wind, from obferving the velocity of the windmill fails; for knowing the radius, and the number of turns in a minute, we fhall have the velocity of the extremities; which, divided by the following divifors, will give the velocity of the wind.

$$
\begin{array}{ll}
\text{Dutch fails in the common pofition} & \begin{cases} \text{unloaded } 4.2 \\ \text{loaded } -3.3 \end{cases} \\[1em]
\text{Dutch fails in their beft pofition} \;\; - & \begin{cases} \text{unloaded } 4.0 \\ \text{loaded } -2.7 \end{cases} \\[1em]
\text{Enlarged fails in their beft pofition} & \begin{cases} \text{unloaded } 4.0 \\ \text{loaded } -2.6 \end{cases}
\end{array}
$$

From the above divifors there arife the following compendiums; fuppofing the radius to be 30 feet, which is the moft ufual length in this country, and the mill to be loaded to a *maximum*, as is ufually the cafe with corn mills; for every *3 turns in a minute, of the Dutch sails in their common position, the wind will move at the rate of 2 miles an hour;* for every *5 turns in a minute, of the Dutch sails in their best position, the wind moves 4 miles an hour;* and for every *6 turns in a minute, of the enlarged sails in their best position, the wind will move 5 miles an hour.*

The following table was communicated to me by my friend Mr. Roufe, and appears to have been conftructed with great care, from a confiderable number of facts and experiments; and having relation to the fubject of this article, I here infert it as he fent it to me: but at the fame time muft obferve, that the evidence for thofe numbers where the velocity of the wind exceeds 50 miles an hour, does not feem of equal authority with that of 50 miles an hour and under. It is alfo to be obferved, that the numbers in column 3. are calculated according to the fquare of the velocity of the wind, which, in moderate velocities, from what has been before obferved, will hold very nearly.

TABLE

TABLE VI.

Containing the Velocity and Force of Wind, according to their common Appellations.

Velocity of the Wind.		Perpendicular force on one foot area in pounds avoirdupois.	Common appellations of the force of winds.
Miles in one Hour.	Feet in one fecond.		
1	1,47	,005	Hardly perceptible.
2	2,93	,020	} Juft perceptible.
3	4,40	,044	
4	5,87	,079	} Gentle pleafant wind.
5	7,33	,123	
10	14,67	,492	} Pleafant brifk gale.
15	22,00	1,107	
20	29,34	1,968	} Very brifk.
25	36,67	3,075	
30	44,01	4,429	} High winds.
35	51,34	6,027	
40	58,68	7,873	} Very high.
45	66,01	9,963	
50	73,35	12,300	A ftorm or tempeft.
60	88,02	17,715	A great ftorm.
80	117,36	31,490	An hurricane.
100	146,70	49,200	An hurricane that tears up trees, carries buildings before it, &c.
1	2	3	

VII. Concerning the abfolute Effect, produced by a given Velocity of the Wind, upon Sails of a given Magnitude and Conftruction.

It has been obferved by practitioners, that in mills with Dutch fails in the common pofition, when they make about 13 turns in a minute, they then work at a mean

L   rate :

rate: that is, by the compendiums in the laft article, when the velocity of the wind is $8\frac{2}{3}$ miles an hour, or $12\frac{2}{3}$ feet in a fecond; which, in common phrafe, would be called a *frefh gale*.

The experiments fet down in Table IV. No. 4, were tried with a wind, whofe velocity was $8\frac{1}{4}$ feet in a fecond; confequently had thofe experiments been tried with a wind, whofe velocity was $12\frac{2}{3}$ feet in a fecond, the effect, by maxim 3d, would have been 3. times greater; becaufe the cube of $12\frac{2}{3}$ is 3 times greater than that of $8\frac{1}{4}$.

From Table IV. No. 4, we find that the fails, when the velocity of the wind was $8\frac{1}{4}$ feet in a fecond, made 130 revolutions in a minute, with a load of 17,52 lb. From the meafures of the machine, preceding the fpecimen of a fet of experiments, we find, that 20 revolutions of the fails raifed the fcale and weight 11,3 inches: 130 revolutions will therefore raife the fcale 73,45 inches, which, multiplied by 17,52 lb. makes a product of 1287, for the effect of the Dutch fails in their beft pofition; that is, when the velocity of the wind is $8\frac{1}{4}$ feet in a fecond: this product therefore multiplied by three, will give 3861 for the effect of the fame fails, when the velocity of the wind is $12\frac{2}{3}$ feet in a fecond.

Defaguliers makes the utmoft power of a man, when working fo as to be able to hold it for fome hours, to be equal to that of raifing an hogfhead of water 10 feet high in a minute. Now, an hogfhead confifting of 63 ale gallons, being reduced into pounds avoirdupois, and the height into inches; the product made by multiplying thofe two numbers will be 76800; which is 19 times greater than the product of the fails laft mentioned, at $12\frac{1}{2}$ feet in a fecond: therefore, by maxim 8th, if we multiply the fquare root of 19, that is 4,46, by 21 inches, the length of the fail producing the effect 3861, we fhall have 93,66 inches, or 7 feet $9\frac{3}{4}$ inches for the radius of a Dutch fail in its beft pofition, whofe mean power fhall be equal to that of a man: but if they are in their common pofition, their length muft be increafed in the ratio of the fquare root of 442 to that of 639, as thus appears;

The ratio of the *maximum* products of No. 8 and 11, Table III. are as 442 : 639: but by maxim 8, the effects of fails of different radii are as the fquare of the radii, confequently the fquare roots of the products or effects, are as the radii fimply; and therefore as the fquare root of 442 is to that of 639; fo is 93,66 to 112,66; or 9 feet $4\frac{3}{4}$ inches.

If the fails are of the enlarged kind, then from Table III. No. 11 and 15, we shall have the square root of 820 to that of 639 : : 93,66 : 82,8 inches, or 6 feet 10⅔ inches: so that in round numbers we shall have the radius of a fail, of similar figure to their respective models, whose mean power shall be equal to that of a man;

The Dutch fails in their common position	- 9½ feet.
The Dutch fails in their best position -	- 8
The enlarged fails in their best position -	- 7

Suppose now the radius of a fail to be 30 feet, and to be constructed upon the model of the enlarged fails, No. 14 or 15, Table III. dividing 30 by 7, we shall have 4,28, the square of which is 18,3 ; and this, according to maxim 7, will be the relative power of a fail of 30 feet, to one of 7 feet ; that is, when working at a mean rate, the 30 feet fail will be equal to the power of 18,3 men, or of 3⅗ horses ; reckoning 5 men to a horse: whereas the effect of the common Dutch fails, of the same length, being less in the proportion of 820 : 442, will be scarce equal to the power of 10 men, or of two horses.

That these computations are not merely speculative, but will nearly hold good when applied to works in large, I have had an opportunity of verifying: for in a mill with the enlarged fails of 30 feet, applied to the crushing of rape seed, by means of two runners upon the edge, for making oil; I observed, that when the fails made 11 turns in a minute, in which case the velocity of the wind was about 13 feet in a second, according to article 6th, the runners then made 7 turns in a minute: whereas 2 horses, applied to the same 2 runners, scarcely worked them at the rate of 3½ turns in the same time. Lastly, with regard to the real superiority of the enlarged fails, above the Dutch fails as commonly made, it has sufficiently appeared, not only in those cases where they have been applied to new mills, but where they have been substituted in the place of the others.

VIII. Concerning horizontal Windmills and Water-wheels, with oblique Vanes.

Observations upon the effects of common windmills, with oblique vanes, have led many to imagine, that could the vanes be brought to receive the direct impulse, like a ship sailing before the wind, it would be a very great improvement in point of power: while others attending to the extraordinary and even unexpected effects of

oblique

oblique vanes have been led to imagine that oblique vanes applied to water-mills, would as much exceed the common water-wheels, as the vertical wind-mills are found to have exceeded all attempts towards an horizontal one. Both thefe notions, but efpecially the firft, have fo plaufible an appearance, that of late years there has feldom been wanting thofe, who have affiduoufly employed themfelves to bring to bear defigns of this kind : it may not therefore be unacceptable to endeavour to fet this matter in a clear light.

PLATE VI. fig. 2. Let A B be the fection of a plane, upon which let the wind blow in the direction C D, with fuch a velocity as to defcribe a given fpace B E, in a given time (fuppofe 1 fecond); and let A B be moved parallel to itfelf, in the direction C D. Now, if the plane A B moves with the fame velocity as the wind; that is, if the point B moves through the fpace B E in the fame time that a particle of air would move through the fame fpace; it is plain that, in this cafe, there can be no preffure or impulfe of the wind upon the plane: but if the plane moves flower than the wind, in the fame direction, fo that the point B may move to F, while a particle of air, fetting out from B at the fame inftant, would move to E, then B F will exprefs the velocity of the plane; and the relative velocity of the wind and plane will be expreffed by the line F E. Let the ratio of F E to B E be given (fuppofe $2 : 3$); let the line A B reprefent the impulfe of the wind upon the plane A B, when acting with its whole velocity B E; but, when acting with its relative velocity F E, let its impulfe be denoted by fome aliquot part of A B, as for inftance $\frac{4}{9}$ A B: then will $\frac{4}{9}$ of the parallelogram A F reprefent the mechanical power of the plane; that is, $\frac{4}{9}$ A B $\times \frac{1}{3}$ B E.

2dly, Let I N be the fection of a plane, inclined in fuch a manner, that the bafe I K of the rectangle triangle I K N may be equal to A B; and the perpendicular N K = B E; let the plane I N be ftruck by the wind, in the direction L M, perpendicular to I K: then, according to the known rule of oblique forces, the impulfe of the wind upon the plane I N, tending to move it according to the direction L M, or N K, will be denoted by the bafe I K; and that part of the impulfe, tending to move it according to the direction I K, will be expreffed by the perpendicular N K. Let the plane I N be moveable in the direction of I K only; that is, the point I in the direction of I K, and the point N in the direction N Q, parallel thereto. Now it is evident, that if the point I moves through the line I K, while a particle of air, fetting forwards at the fame time from the point N, moves through the line N K, they will both arrive at the point K at the fame time; and confequently, in this cafe alfo, there can be no preffure or impulfe of the particle of the air upon the plane I N. Now let I O be to I K as B F to B E; and let

the

the plane I N move at fuch a rate, that the point I may arrive at O, and acquire the pofition I Q, in the fame time that a particle of wind would move through the fpace N K: as O Q is parallel to I N; (by the properties of fimilar triangles) it will cut N K in the point P, in fuch a manner, that N P = B F, and P K = F E: hence it appears, that the plane I N, by acquiring the pofition O Q, withdraws itfelf from the action of the wind, by the fame fpace N P, that the plane A B does by acquiring the pofition F G; and confequently, from the equality of P K to F E, the relative impulfe of the wind P K, upon the plane O Q, will be equal to the relative impulfe of the wind F E, upon the plane F G: and fince the impulfe of the wind upon A B, with the relative velocity F E, in the direction B E, is reprefented by $\frac{4}{9}$ A B; the relative impulfe of the wind upon the plane I N, in the direction N K, will in like manner be reprefented by $\frac{4}{9}$ I K; and the impulfe of the wind upon the plane I N, with the relative velocity P K, in the direction I K, will be reprefented by $\frac{4}{9}$ N K: and confequently the mechanical power of the plane I N, in the direction I K, will be $\frac{4}{9}$ the parallelogram I Q: that is $\frac{1}{7}$ I K $\times \frac{4}{9}$ N K: that is, from the equality of I K = A B and N K = B E, we fhall have $\frac{4}{9}$ I Q = $\frac{1}{7}$ A B $\times \frac{4}{9}$ B E = $\frac{4}{9}$ A B $\times \frac{1}{7}$ B E = $\frac{4}{9}$ of the area of the parallelogram A F. Hence we deduce this

GENERAL PROPOSITION,

That all Planes, however situated, that intercept the same Section of the Wind, and having the same relative Velocity, in regard to the Wind, when reduced into the same Direction, have equal Powers to produce mechanical Effects.

For what is loft by the obliquity of the impulfe, is gained by the velocity of the motion.

Hence it appears, that an oblique fail is under no difadvantage in refpect of power, compared with a direct one; except what arifes from a diminution of its breadth, in refpect to the fection of the wind: the breadth I N being obliquity reduced to I K.

The difadvantage of horizontal windmills therefore does not confift in this; that each fail, when directly expofed to the wind, is capable of a lefs power, than an oblique one of the fame dimenfions; but that in an horizontal windmill, little more than one fail can be acting at once: whereas in the common windmill, all the four act together: and therefore,

therefore, fuppofing each vane of an horizontal windmill, of the fame dimenfions as each vane of the vertical, it is manifeft the power of a vertical mill with four fails, will be four times greater than the power of the horizontal one, let its number of vanes be what it will; this difadvantage arifes from the nature of the thing: but if we confider the further difadvantage, that arifes from the difficulty of getting the fails back again againft the wind, &c. we need not wonder if this kind of mill is in reality found to have not above $\frac{1}{8}$ or $\frac{1}{10}$ of the power of the common fort; as has appeared in fome attempts of this kind.

In like manner, as little improvement is to be expected from water-mills with oblique vanes: for the power of the fame fection of a ftream of water, is not greater when acting upon an oblique vane, than when acting upon a direct one: and any advantage that can be made by intercepting a greater fection, which fometimes may be done in the cafe of an open river, will be counterbalanced by the fuperior refiftance, that fuch vanes would meet with by moving at right angles to the current: whereas the common floats always move with the water nearly in the fame direction.

Here it may reafonably be afked, that fince our geometrical demonftration is general, and proves, that one angle of obliquity is as good as another; why in our experiments it appears, that there is a certain angle which is to be preferred to all the reft? It is to be obferved, that if the breadth of the fail I N is given, the greater the angle K I N, and the lefs will be the bafe I K: that is, the fection of wind interfected, will be lefs: on the other hand, the more acute the angle K I N, the lefs will be the perpendicular K N: that is, the impulfe of the wind, in the direction I K being lefs, and the velocity of the fail greater; the refiftance of the medium will be greater alfo. Hence therefore, as there is a diminution of the fection of the wind intercepted on one hand, and an increafe of refiftance on the other, there is fome angle, where the difadvantage arifing from thefe caufes upon the whole is the leaft of all; but as the difadvantage arifing from refiftance is more of a phyfical than geometrical confideration, the true angle will beft be affigned by experiments.

Scholium.

In trying the experiments contained in Table III. and IV. the different. fpecific gravity of the air, which is undoubtedly different at different times, will caufe a difference in the load, proportional to the difference of its fpecific gravity, though its

velocity

velocity remains the fame; and a variation of fpecific gravity may arife not only from a variation of the weight of the whole column, but alfo by the difference of heat of the air concerned in the experiment, and poffibly of other caufes; yet the irregularities that might arife from a difference of fpecific gravity were thought to be too fmall to be perceivable, till after the principal experiments were made, and their effects compared; from which, as well as fucceeding experiments, thofe variations were found to be capable of producing a fenfible, though no very confiderable effect: however, as all the experiments were tried in the fummer feafon, in the day time, and under cover, we may fuppofe that the principal fource of error would arife from the different weight of the column of the atmofphere at different times: but as this feldom varies above $\frac{1}{15}$ part of the whole, we may conclude, that though many of the irregularities contained in the experiments referred to in the foregoing effay, might arife from this caufe; yet as all the principal conclufions are drawn from the medium of a confiderable number, many whereof were made at different times, it is prefumed that they will nearly agree with the truth, and be altogether fufficient for regulating the practical conftruction of thofe kind of machines, for which ufe they were principally intended.

AN

AN experimental Examination of the Quantity and Proportion of Mechanic Power necessary to be employed in giving different Degrees of Velocity to Heavy Bodies from a State of rest, by Mr. JOHN SMEATON, F. R. S.

Read before the Royal Society, April 25th, 1776.

ABOUT the year 1686, Sir ISAAC NEWTON first published his *Principia*, and, conformably to the language of mathematicians of those times defined, that " the " quantity of motion is the measure of the same, arising from the velocity and quantity " of matter conjointly." Very soon after this publication, the truth or propriety of this definition was disputed by certain philosophers, who contended, that the measure of the quantity of motion should be estimated by taking the quantity of matter and the square of the velocity conjointly. There is nothing more certain, than that from equal impelling powers, acting for equal intervals of time, equal increases of velocity are acquired by given bodies, when unresisted by a medium ; thus gravity causes a body, in obeying its impulse during one second of time, to acquire a velocity which would carry it uniformly forward, without any additional impulse, at the rate of 32 ft. 2 in. *per* second ; and if gravity is suffered to act upon it for two seconds, it will have, in that time, acquired a velocity that would carry it, at an uniform rate, just double of the former ; that is, at the rate of 64 ft. 4 in. *per* second. Now, if in consequence of this equal increase of velocity, in an equal increase of time, by the continuance of the same impelling power, we define that to be a double quantity of motion, which is generated in a given quantity of matter, by the action of the same impelling power for a double time ; this will be co-incident with Sir ISAAC NEWTON's definition above mentioned ; whereas, in trying experiments upon the total effects of bodies in motion, it appears, that when a body is put in motion, by whatever cause, the impression it will make upon an uniformly resisting medium, or upon uniformly yielding substances, will be as the mass of matter of the moving body, multiplied by the square of its velocity : the question, therefore, properly is, whether those terms, the *quantity of motion*, the *momenta* of bodies in motion, or *forces* of bodies in motion, which have generally been esteemed synonymous, are with the most propriety of language to be esteemed equal, double, or triple, when they have been generated by an equable impulse, acting for an equal, double, or triple, time ; or that it should be measured by the effects being equal, double, or triple, in overcoming resistances before a body in motion can be stopped ?

ſtopped? For, according as thoſe terms are underſtood in this or that way, it will neceſſarily follow, that the *momenta* of equal bodies will be as the velocities, or as the ſquares of the velocities, or as the ſquares of the velocities reſpectively; it being certain, that, whichever we take for the proper definition of the term quantity of motion, by paying a proper regard to the collateral circumſtances that attend the application of it, the ſame concluſion, in point of computation, will reſult. I ſhould not, therefore, have thought it worth while to trouble the Society upon this ſubject, had I not found, that not only myſelf and other practical artiſts, but alſo ſome of the moſt approved writers, had been liable to fall into errors, in applying theſe doctrines to practical mechanics, by ſometimes forgetting or neglecting the due regard which ought to be had to theſe collateral circumſtances. Some of theſe errors are not only very conſiderable in themſelves, but alſo of great conſequence to the public, as they tend greatly to miſlead the practical artiſt in works that occur daily, and which often require very great ſums of money in their execution. I ſhall mention the following inſtances.

DESAGULIERS, in his ſecond volume of Experimental Philoſophy, treating upon the queſtion concerning the forces of bodies in motion, after taking much pains to ſhew that the diſpute, which had then ſubſiſted fifty years, was a diſpute about the meaning of words; and that the ſame concluſion will be brought out, when things are rightly underſtood, either upon the old or new opinion, as he diſtinguiſhes them; among other things, tells us, that the old and new opinion may be eaſily reconciled in this inſtance: that the wheel of an underſhot water-mill is capable of doing quadruple work when the velocity of the water is doubled, inſtead of double work only; " becauſe (the adjutage " being the ſame), ſays he, we find, that as the water's volocity is double, there are " twice the number of particles of water that iſſue out, and therefore the ladle-board is " ſtruck by twice the matter, which matter moving with twice the velocity that it had " in the firſt caſe, the whole effect muſt be quadruple, though the inſtantaneous ſtroke " of each particle is increaſed only in a ſimple proportion of the velocity." See vol. II. Annotations on lecture 6th, p. 92.

Again, in the ſame volume, lecture 12th, p. 424, referring to what went before, he tells us, " The knowledge of the foregoing particulars is abſolutely neceſſary for ſetting " an underſhot wheel to work: but the advantage to be reaped from it would be ſtill " gueſswork, and we ſhould be ſtill at a loſs to find out the utmoſt it can perform, if " we had not an ingenious propoſition of that excellent mechanic M. PARENT, of the " Royal Academy of Sciences, who has given us a *maximum* in this caſe, by ſhewing,

M   " that

" that an undershot wheel can do the most work, when its velocity is equal to the third
" part of the velocity of the water that drives it, &c. because then two-thirds of the
" water is employed in driving the wheel with a force proportionable to the square of its
" velocity. If we multiply the surface of the adjutage or opening by the height of the
" water, we shall have the column of water that moves the wheel. The wheel thus
" moved will sustain on the opposite side only four-ninths of that weight, which will
" keep it in equilibrio; but what it can move with the velocity it goes with, will be
" but one-third of that weight of equilibrium; that is, $\frac{4}{27}$ths of the weight of the first
" column, &c.—This is the utmost that can be expected."

The same conclusion is likewise adopted by MACLAURIN, in art. 907. p. 728. of his
Fluxions, where, giving the fluxionary deduction of M. PARENT's proposition, he says,
" that if A represents the weight which would balance the force of the stream, when its
" velocity is a; and υ represents the velocity of the part of the engine, which it strikes
" when the motion of the machine is uniform, &c.—the machine will have the greatest
" effect when υ is equal to $\dfrac{a}{3}$; that is, if the weight that is raised by the engine be less
" than the weight which would balance the power, in the proportion of 4 to 9, and the
" *momentum* of the weight is $\dfrac{4\text{A}a}{27}$"

Finding that these conclusions were far from the truth; and seeing, from many other
circumstances, that the practical theory of making water and wind-mills was but very
imperfectly delivered by any author I had then an opportunity of consulting *; in the

---

* BELIDOR, *Architecture Hydraulique*, greatly prefers the application of water to an undershot mill,
instead of an overshot; and attempts to demonstrate, that water applied undershot will do six times
more execution than the same applied overshot. See vol. I. p. 286. While DESAGULIERS, endeavouring
to invalidate what had been advanced by BELIDOR, and greatly preferring an overshot to an undershot,
says, Annotat. on lecture 12, vol. II. p. 532, that from his own experience, " a well-made overshot mill
" ground as much corn in the same time with ten times less water;" so that betwixt BELIDOR and
DESAGULIERS, here is a difference of no less than 60 to 1.

Again, BELIDOR, vol. II. p. 72, says, that the centre of gravity of each sail of a wind-mill should
travel in its own circle with one-third of the velocity of the wind; so that, taking the distance of this
centre of gravity from the centre of motion at 20 feet, as he states it, p. 38, article 849, the circumference will be exceeding 126 feet English measure: a wind, therefore, to make the mill go twenty
turns *per* minute, which they frequently do with a fresh wind and all their cloth spread, would require
the wind to move above 80 miles an hour; a velocity perhaps hardly equalled in the greatest storms we
experience in this climate.

year 1751 I began a courfe of experiments upon this fubject. Thefe experiments, with the conclufions drawn from them, have already been communicated to this Society, who printed them in vol. LI. of their Tranfactions for the year 1759, and for this communication I had the honour of receiving the annual medal of Sir GODFREY COPLEY, from the hands of our very worthy Prefident the late Earl of MACCLESFIELD. Thofe experiments and conclufions ftand uncontroverted, fo far as I know, to this day; and having fince that time been concerned in directing the conftruction of a great number of mills, which were all executed upon the principles deduced from them, I have by that means had many opportunities of comparing the effects actually produced with the effects which might be expected from the calculation; and the agreement, I have always found between thefe two, appears to me fully to eftablifh the truth of the principles upon which they were conftructed, when applied to great works, as well as upon a fmaller fcale in models.

Refpecting the explanatory deduction of DESAGULIERS in the firft example above-mentioned, which, indeed, I have found to be the commonly received doctrine among theoretical mechanics, it is fhewn above, in my former Effay, page 41, &c. part 1. maxim 4. that, where the velocity of water is double, the adjutage or aperture of the fluice remaining the fame, the effect is eight times; that is, not as the fquare but as the cube of the velocity; and the fame is inveftigated concerning the power of the wind arifing from difference of velocity, p. 67. being part 3. maxim 4.

The conclufion in the fecond example above-mentioned, adopted both by DESAGULIERS and MACLAURIN, is not lefs wide of the truth than the foregoing; for if that conclufion were true, only $\frac{4}{27}$ths of the water expended could be raifed back again to the height of the refervoir from which it had defcended, exclufively of all kinds of friction, &c. which would make the actual quantity raifed back again ftill lefs; that is, lefs than one-feventh of the whole; whereas it appears, from Table I. of the preceding effay, that in fome of the experiments there related, even upon the fmall fcale on which they were tried, the work done was equivalent to the raifing back again about one quarter of the water expended; and in large works the effect is ftill greater, approaching towards half, which feems to be the limit for the underfhot mills, as the whole would be the limit for the overfhot mills, if it were poffible to fet afide all friction, refiftance from the air, &c.

The velocity alfo of the wheel, which, according to M. PARENT's determination, adopted by DESAGULIERS and MACLAURIN, ought to be no more than one-third of that

of

of the water, varies at the *maximum* in the abovementioned experiments of Table I. between one-third and one-half; but in all the cafes there related, in which the moft work is performed in proportion to the water expended, and which approach the neareft to the circumftances of great works, when properly executed, the *maximum* lies much nearer to one-half than one-third; one-half, feeming to be the true *maximum*, if nothing were loft by the refiftance of the air, the fcattering of the water carried up by the wheel, and thrown off by the centrifugal force, &c. all which tend to diminifh the effect more, at what would be the *maximum* if thefe did not take place, than they do when the motion is a little flower.

Finding thefe matters, as well as others, to come out in the experiments, fo very different from the opinions and calculations of authors of the firft reputation, who, reafoning according to the Newtonian definition, muft have been led into thefe errors from a want of attending to the proper collateral circumftances; I thought it very material, efpecially for the praftical artift, that he fhould make ufe of a kind of reafoning in which he fhould not be fo liable to miftakes; in order, therefore, to make this matter perfeftly clear to myfelf, and poffibly fo to others, I refolved to try a fet of experiments from whence it might be inferred, what proportion or quantity of mechanical power is expended in giving the fame body different degrees of velocity. This fcheme was put in execution in the year 1759, and the experiments were then fhewn to feveral friends, particularly my very worthy and ingenious friend Mr. WILLIAM RUSSELL.

In my experimental inquiry concerning the powers of water and wind before referred to, I have, p. 30, part 1, defined what I meant by power, as applied to praftical mechanics, that is, what I now call mechanical power; which, in terms fynonymous to thofe there ufed, may be faid to be meafured by multiplying the weight of the body into the perpendicular height from which it can defcend; thus the fame weight, defcending from a double height, is capable of producing a double mechanical effeft, and is therefore a double mechanical power. A double weight defcending from the fame height is alfo a double power, becaufe it likewife is capable of producing a double effeft; and a given body, defcending through a given perpendicular height, is the fame power as a double body defcending through half that perpendicular; for, by the intervention of proper levers, they will counter-balance one another, conformably to the known laws of mechanics, which have never been controverted. It muft, however, be always underftood, that the defcending body, when afting as a meafure of

Plate 7.page 85.

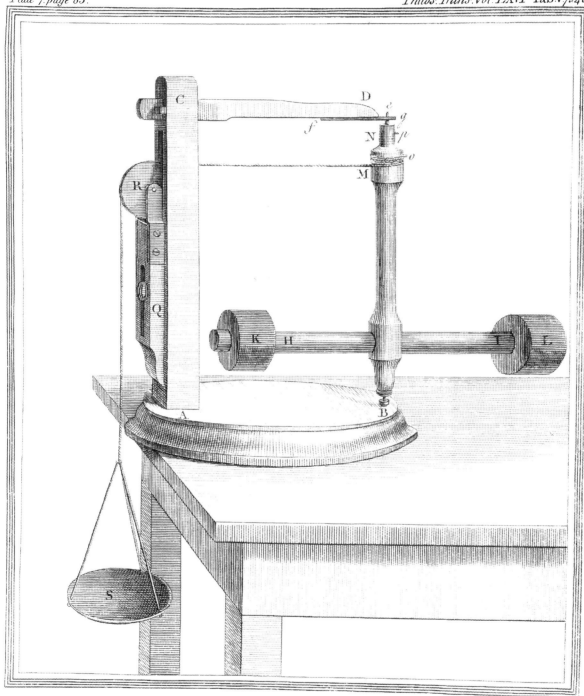

power, is fuppofed to defcend flowly, like the weight of a clock or a jack; for, if quickly defcending, it is fenfibly compounded with another law, viz. the law of acceleration by gravity.

DESCRIPTION OF THE MACHINE.

PLATE VI.

A B is the bafe of the machine placed upon a table.

A C is a pillar or ftandard.

C D is an arm, upon the extremity of which is fixed a plate $f g$, which is here feen edge-ways, through which is a fmall hole for receiving a fmall fteel pivot e, fixed in the top of the upright axis e B; the lower end of this axis finifhes in a conical fteel point, which refts upon a fmall cup of hard fteel polifhed at B.

H I is a cylinder of white fir, which paffes through a perforation in the axis, and therein fixes; and, upon the two arms formed thereby, are capable of fliding.

K L two cylindric weights of lead of equal fize, which are capable of being fixed upon any part of the cylindric arms, from the axis to their extremities, by means of two thin wedges of wood. The two weights, therefore, being at equal diftances from the centre, and the axis perpendicular, the whole will be balanced upon the point at B, and moveable thereupon by an impelling power with very little friction.

Upon the upper part of the axis are formed M N, two cylindrical barrels, whereof M is double the diameter of N: they have a little pin ftuck into one fide of each at o, p.

Q is a piece capable of fliding higher or lower, as occafion requires; and carries

R, a light pulley of about three inches diameter, hung upon a fteel axis, and moveable upon two fmall pivots. The plane of the pulley, however, is not directed to the middle of the upright axis, but a little on one fide, fo as to point (at a mean) between the furface of the bigger barrel and the lefs.

S is

s is a light fcale for receiving weights, and hangs by a fmall twine, cord, or line, that paffes the pulley and terminates either upon the bigger barrel or the lefs, as may be required ; the fliding-piece ϙ being moved higher or lower for each, that the line, in paffing from the pulley to the barrel, may be nearly horizontal. The end of the line that is furtheft from the fcale is terminated by a fmall loop, which hangs on upon the pin *o*, or the pin *p*, according as the bigger or the leffer barrel is to be ufed.

Now, having wound up a certain number of turns of the line upon the barrel, and having placed a weight in the fcale s, it is obvious, that it will caufe the axis to turn round, and give motion to its arms, and to the weights of lead placed thereon, which are the heavy bodies to be put in motion by the impulfe of the weight in the fcale ; and when the line is wound off to the pin, the loop flips off, and the fcale then falling down, the weight will ceafe to accelerate the motion of the heavy bodies, and leave them revolving, equably forward, with the velocity they have acquired, except fo far as it muft be gradually leffened by the friction of the machine and refiftance of the air, which being fmall, the bodies will revolve fometime before their velocity is apparently diminfhed.

Measures of some Parts of the Machine.

	Inches.
Diameter of the cylinders of lead, or the heavy bodies, -	2,57
Length of ditto, - - - - - - - -	1,56
Diameter of the hole therein, - - - - -	,72

Weight of each cylinder 3 lbs. Avoirdupois.

Greater diftance of the middle of each body from the centre of the axis, - - - - - -	8,25
The fmaller diftance of ditto, - - - - -	3,92
10 turns of the fmaller barrel raifes the fcale, 5 ditto of the bigger ditto, - - - - - - -	25,25

When the bodies are at the fmaller diftance above fpecified from the axis of rotation, they are then in effect at half the greater diftance from that axis ; for, fince the axis

itfelf,

itfelf, and the cylindric arms of wood, keep an unvaried diftance from the centre of rotation, the bodies themfelves muft be moved nearer than half their former diftance, in order that, compounded with the unvariable parts, they may be virtually at the half diftance. In order to find this half diftance nearly, I put in an arm of the fame wood, that only went through the axis, without extending in the oppofite direction; one of the bodies being put upon the end of this arm, at the diftance of 8,25 inches, the whole machine was inclined till the body and arm became a kind of pendulum, and vibrated at the rate of 92 times *per* minute; and as a pendulum of the half length vibrates quicker in the proportion of $\sqrt{1}$ to $\sqrt{2}$; that is, in the proportion of 92 to 130 nearly; therefore, keeping the fame inclination of the machine, the weight was moved upon the arm till it made 130 vibrations *per* minute; which was found to be, when it was at 3,92 inches diftance from the centre as above ftated, which is about $\frac{2}{10}$ths of an inch nearer than the half diftance. The double arm was then put in, and marked accordingly, and the bodies being mounted thereon, the whole was adjufted ready for ufe; and with it were tried the following experiments, each of which was repeated fo many times as to be fully fatisfactory.

Table of Experiments.

Number.	Ounces Avoirdu-pois in the Scale.	Barrel ufed, M the bigger, N the fmaller.	The Arms, W the whole, H the half-length.	Number of Turns of the Line wound on the Barrel.	Time of the De-fcent of the Weight in the Scale.	Time in making 20 Revolutions with equal Motions.
					″	′
1	8	M	W	5	$14\frac{1}{4}$	29
2	8	N	W	10	$28\frac{1}{4}$	$29\frac{1}{4}$
3	8	N	W	$2\frac{1}{2}$	$14\frac{1}{4}$	$58\frac{1}{2}$
4	32	M	W	5	7	14
5	32	N	W	10	14	$14\frac{3}{4}$
6	32	N	W	$2\frac{1}{2}$	7	$28\frac{3}{4}$
7	8	M	H	5	7	$14\frac{3}{4}$
8	8	N	H	10	14	15
9	8	N	H	$2\frac{1}{2}$	7	$30\frac{1}{4}$
1	2	3	4	5	6	7

The

The $58''\frac{1}{2}$ in number 3, column 7. was determined in fact from $29'\frac{1}{4}$, being the time of making 10 equable revolutions after the weight was dropped off, in order to prevent the fenfible retardation that might take place, and affect the obfervation, if continued for 20 revolutions made fo flowly.

Further Definitions.

I have already defined what I mean by mechanic power; but, before I proceed further, it will be neceffary alfo to define the following terms:

Impulfe or Impulfion, } By all which, I underftand the uniform endeavour that
Impulfive Force or Power, } one body exerts upon another, in order to make it move;
Impelling Force or Power, } and that, whether it produces or generates motion by this endeavour or not; and the quantity of this impelling power may be meafured either by its being a weight of itfelf, or by being counterbalanced by a weight. It may alfo act either immediately upon the body to be moved, fo that if motion is the confequence, they move with the fame velocity; and that, either by a fimple contact, or by being drawn as by a cord, or pufhed as by a ftaff: or it may act by the intervention of a lever or other mechanic inftrument, in which the velocity of the body to be moved may be very different from the velocity of the impelling power or mover; but in comparing them, the impelling powers muft be reduced according to the proportional velocities of the mover and moved; or, in levers of different lengths, they may be compared by a ftandard length of lever, which is the method taken in the fubfequent reafoning upon the preceding experiments. An impelling power, therefore, confifting of a double weight, or requiring a double weight to counter-balance it, when acting with equal levers, is a double impelling power, or an impelling power of double the intenfity.

Observations and Deductions from the preceding Experiments.

1ft, By the firft experiment it appears, that the mechanic power employed, confifting of 8 ounces in the fcale, deliberately defcending (by 5 turns of the bigger barrel) through a perpendicular fpace $25\frac{1}{4}$ inches, will reprefent the quantity of mechanic power which caufes the two heavy bodies, from a ftate of reft, to acquire a velocity, fuch as to carry them equably through 20 circumferences of their circle of revolution in the fpace of $29''$; and that the time in which the mechanic power produced this effect was $14''\frac{1}{4}$, as appears by column 6th. And this mechanic power we fhall exprefs by the number

number 202, the product of the number of ounces in the fcale multiplied by the inches in its perpendicular defcent, for $8 \times 25\frac{1}{4} = 202$.

2dly, By the fecond experiment, as 10 turns of the fmaller barrel are equal to the fame perpendicular height as 5 turns of the bigger, it follows, that the fame mechanic power, viz. 202, acting upon the fame heavy bodies to accelerate them, produces the very fame effect in generating motion in the bodies as it did before, viz. 20 revolutions in $29''\frac{1}{4}$, the fmall difference of $\frac{1}{4}$ of a fecond being no more than may reafonably be attributed to the unavoidable errors arifing from friction of the machine, want of per-fect accuracy in its meafures, refiftance of the air, and imperfections in the obferva-tions themfelves, which muft not only be allowed for in this, but the reft; but as the impelling power is acting here upon a lever of but half the length, and, confequently, but half the intenfity, when referred to the bodies to be moved, it takes juft double the time to generate the fame velocity therein.

DEDUCTION. It appears from hence, that the fame mechanic power is capable of producing the fame velocity in a given body, whether it is applied fo as to produce it in a greater or leffer time; but that the time taken to produce a given velocity, by an uniformly continued action, is in a fimple inverfe proportion of the intenfity of the impulfive power.

3dly, The third experiment being made with $2\frac{1}{2}$ turns of the leffer barrel, the fame weight in the fcale of 8 ounces defcending only one quarter part of the former perpendicular, the mechanic power employed will be only one quarter part of the former, viz. $50\frac{1}{2}$; but as one quarter part of the mechanic power produces half of the former velocity in the heavy bodies; that is, they make 20 revolutions in $58''\frac{1}{2}$; that is, nearly 10 revolutions in $29''$; we may conclude, in this inftance, that the mechanic power, employed in producing motion, is as the fquare of the velocity produced in the fame body; and that the velocity produced is as the time that an impelling power, of the fame intenfity, continues to act upon it, as appears by the near agreement of numbers 2 and 3, column 6th.

4thly, In the fourth experiment, the apparatus is the fame as the firft, only here the weight in the fcale is 32 ounces; that is, the impelling power is the quadruple of the firft, and hereby a double velocity is given to the bodies; for they make 20 re-volutions in $14''$, which is a fmall matter lefs than half the time taken up in making 20 revolutions in the firft experiment. It alfo appears, that the velocity acquired is

N

fimply

fimply as the impelling power compounded with the time of its action : for a quadruple impulfion acting for 7″ inftead of 14″ generates a double velocity, while the mechanic power employed to generate it is quadruple, for $32 \times 25\frac{1}{4} = 808$. And here the mechanic power employed being four times greater than the firft, it holds here alfo, that the mechanic power, to be neceffarily employed, is as the fquare of the velocity to be generated; that is, in the fame proportion as turned out in the third experiment, where the mechanic power employed was only a quarter part of the firft.

5thly, The fifth and fixth experiments were made with a mechanic power four times greater than thofe employed in numbers 2 and 3 refpectively; and fince the fame deductions refult from hence as from numbers 2 and 3, they are additional confirmations of the conclufions drawn from them and from the laft article.

6thly, In the feventh experiment, the difpofition of the apparatus is the fame as number 1, only here the bodies are placed upon the arms at the half-length; from whence it appears, that the fame mechanic power ftill produces the fame velocity in the fame bodies; for though 20 revolutions were performed in $14\frac{3}{4}$″ (fee column 7) which is nearly half the time that 20 revolutions were performed in the firft experiment; yet, fince the circles in which the bodies revolved in the feventh are only of half the circumference as thofe of number 1, it is obvious, that the abfolute velocity acquired by the moving bodies in the two cafes is equal. But, by column 6th, the time in which it was generated is only half; yet, notwithftanding, this will coincide with the former conclufions, if the intenfity of the impelling power is compounded therewith; for, though the barrel was the fame with the fame number of turns as in number 1, and therefore the lever the fame, by which the impelling power acted, yet, as the bodies, upon which this lever was to act, were placed upon a lever of only half the length from the centre, the impelling power acting by the firft lever, would act upon the fecond with double the intenfity, according to the known laws of mechanics; that is, it would require a double weight oppofing the bodies, to prevent their moving, in order to balance it. An impulfive power, therefore, of double the intenfity, acting for half the time, produces the fame effect in generating motion, as an impulfive power, of half the intenfity, acting for the whole time.

7thly, The eighth and ninth experiments afford the fame deductions and confirmations relative to the feventh experiment, that the fifth and fixth do refpecting the fourth, and that the fecond and third do refpecting the firft; and from the near agreement

ment

ment of the whole, when the neceſſary allowances before mentioned are made, together with ſome ſmall inequality ariſing from the mechanical power loſt by the difference of the motion given by gravity to the weight in the ſcale: I ſay, from theſe agreements, under the very different mechanical powers applied, which were varied in the proportion of 1 to 16, we may ſafely conclude, that this is the univerſal law of nature, reſpecting the capacities of bodies in motion to produce mechanical effects, and the quantity of mechanic power neceſſary to be employed to produce or generate different velocities (the bodies being ſuppoſed equal in their quantity of matter); that the mechanic powers to be expended are as the ſquares of the velocities to be generated, and *vice verſâ*; and that the ſimple velocities generated are as the impelling power compounded with, or multiplied by, the time of its action, and *vice verſâ*.

We ſhall, perhaps, form a ſtill clearer conception of the relation between velocities produced, and the quantity of mechanic power required to produce them; together with the collateral circumſtances attending, by which theſe propoſitions. ſeemingly two, are reconciled and united, by ſtating the following popular elucidation, which indeed was the original idea that occurred to me on conſidering this ſubject; to put which to an experimental proof gave birth to the foregoing apparatus and experiments.

Suppoſe then a large iron ball of 10 feet diameter, turned truly ſpherical, and ſet upon an extended plane of the ſame metal, and truly level. Now, if a man begins to puſh at it, he will find it very reſiſting to motion at firſt; but, by continuing the impulſe, he will gradually get it into motion, and having nothing to reſiſt it but the air, he will, by continuing his efforts, at length get it to roll almoſt as faſt as he can run. Suppoſe now, in the firſt minute he gets it rolled through a ſpace of one yard; by this motion, proceeding from reſt (ſimilar to what happens to falling bodies) it would continue to roll forward at the rate of two yards *per* minute, without further help; but ſuppoſing him to continue his endeavours, at the end of another minute he will have given it a velocity capable of carrying it through a ſpace of two yards more, in addition to the former, that is, at the rate of four yards *per* minute; and at the end of the third minute, he has again added an equal increaſe of velocity, and made it proceed at the rate of ſix yards *per* minute; and ſo on, increaſing its velocity at the rate of two yards in every minute. The man, therefore, in the ſpace of every minute exerts an equal impulſe upon the ball, and generates an equal increaſe of movement correſpondent to the definition of Sir ISAAC NEWTON. But let us ſee what happens beſides: the man, in the firſt minute, has moved but one yard from where he ſet out; but he muſt in the fecond minute move two yards more, in order to keep up with the ball; and as he ex-erted an impulfe upon it, fo as at the end of the fecond minute to have given it an ad-ditional velocity of the two yards, he muft alfo in this time have gradually changed its velocity from the rate of two yards *per* minute to that of four, and the fpace, that he will of confequence have actually been obliged to go through in the fecond minute, will be according to the mean of the extremes of velocity at the beginning and end thereof, that is, three yards in the fecond minute; fo that being one yard from his original place at the begining of the fecond minute, at the end of it he will have moved the fum of the journies of the firft and fecond minute, that is, in the whole four yards from his origi-nal place. As he has now generated a velocity in the ball of four yards *per* minute, the third minute he muft travel four yards to keep up with the ball, and one more in generating the equal increment of velocity; fo that in the third minute, he muft travel five yards to keep up the fame impelling power upon the ball that he did in the firft minute in travelling one, fo that thefe five yards in the third minute, added to the four yards that he had travelled in the two preceding minutes, fets him at the end of the third minute nine yards from whence he fet out, having then given the ball a velocity capable of carrying it uniformly forward at the rate of fix yards *per* minute, as before ftated. We may now leave the further purfuit of thefe proportions, and fee how the account ftands. He generated a velocity of two yards *per* minute in the firft minute, the fquare of which is four, when he had moved but one yard from his place; and he had generated a velocity of fix yards *per* minute, the fquare of which is thirty-fix, at the end of the third minute, when he had travelled nine yards from his place. Now, fince the fquare of the velocity, generated at the end of the firft minute, is to that of the velocity generated at the end of the third minute, as 4 : 36, that is, as 1 : 9; and fince the fpaces, moved through by the man to communicate thefe velocities, are alfo as 1 : 9, it follows, that the fpaces through which the man muft travel, in order to generate thefe velocities refpectively (preferving the impelling power perfectly equal), muft be as the fquares of the velocities com-municated to the ball; for, if the man was to be brought back again to his original place by a mechanical power equally exerted upon the man equally refifting, this would be the meafure of what the man has done in order to give a motion to the ball. It there-fore directly follows, conformably to what has been deduced from the experiments, that the mechanic power that muft of neceffity be employed in giving different degrees of velocity to the fame body, muft be as the fquare of that velocity; and if the converfe of this propofition did not hold, viz. that if a body in motion, in being ftopped, would not produce a mechanical effect equal or proportional to the fquare of its velocity, or to the mechanical power employed in producing it, the effect would not correfpond with its producing caufe.

Thus

Thus the confequences of generating motion upon a level plane exactly correfpond with the generating of motion by gravity; viz. that though in two feconds of time the equal impulfive power of gravity gives twice the velocity to a body that it does in one fecond, yet this collateral circumftance attends it, that at the end of the double time, in confequence of the velocity acquired in the firft half, the body has fallen from where it fet forward through four times the perpendicular; and therefore, though the velocity is only doubled, yet four times the mechanical power has been confumed in producing it, as four times the mechanical power muft be expended in bringing up the fallen body to its firft place.

This then appers to be the foundation, not only of the difputes that have arifen, but of the miftakes that have been made, in the application of the different definitions of quantity of motion; that while thofe that have adhered to the definition of Sir ISAAC NEWTON, have complained of their adverfaries, in not confidering the time in which effects are produced, they themfelves have not always taken into the account the fpace that the impelling power is obliged to travel through, in producing the different degrees of velocity. It feems, therefore, that, without taking in the collateral circumftances both of time and fpace, the terms, quantity of motion, *momentum*, and force of bodies in motion, are abfolutely indefinite; and that they cannot be fo eafily, diftinctly, and fundamentally compared, as by having recourfe to the common meafure, viz. mechanic power.

From the whole of what has been inveftigated, it therefore appears, that time, properly fpeaking, has nothing to do with the production of mechanical effects, otherwife than as, by equally flowing, it becomes a common meafure; fo that, whatever mechanical effect is found to be produced in a given time, the uniform continuance of the action of the fame mechanical power will, in a double time, produce two fuch effects, or twice that effect. A mechanical power, therefore, properly fpeaking, is meafured by the whole of its mechanical effect produced, whether that effect is produced in a greater or a leffer time; thus, having treafured up 1000 tuns of water, which I can let out upon the overfhot wheel of a mill, and defcending through a perpendicular of 20 feet, this power applied to proper mechanic inftruments, will produce a certain effect, that is, it will grind a certain quantity of corn; and that at a certain rate of expending it, it will grind this corn in an hour. But fuppofe the mill equally adapted to produce a proportionable effect, by the application of a greater impulfive power as with a lefs; then, if I let out the water twice as faft upon the wheel, it will grind the corn twice as faft, and both the water will be expended and the corn ground in half an hour. Here the fame mechanical effect is

produced;

produced; viz. the grinding a given quantity of corn, by the fame mechanical power, viz. 1000 tuns of water defcending through a given perpendicular of 20 feet, and yet this effect is in one cafe produced in half the time of the other. What time, therefore, has to do in the bufinefs is this : let the rate of doing the bufinefs, or producing the effect, be what it will, if this rate is uniform, when I have found by experiment what is done in a given time, then, proceeding at the fame rate, twice the effect will be produced in twice the time, on fuppofition that I have a fupply of mechanic power to go on with. Thus 1000 tuns of water, defcending through 20 feet of perpendicular, being, as has been fhewn, a given mechanic power, let me expend it at what rate I will, if when this is expended, I muft wait another hour before it be renewed, by the natural flow of a river, or otherwife, I can then only expend twelve fuch quantities of power in 24 hours; but if, while I am expending 1000 tuns in one hour, the ftream renews me the fame quantity, then I can expend 24 fuch quantities of power in 24 hours; that is, I can go on continually at that rate, and the product or effect will be in proportion to time, which is the common meafure; but the quantity of mechanic power arifing from the flow of the two rivers, compared by taking an equal portion of time, is double in the one to the other, though each has a mill, that, when going, will grind an equal quantity of corn in an hour.

NEW Fundamental Experiments upon the Collision of Bodies, by Mr. JOHN SMEATON, F. R. S.

Read before the Royal Society, April 18th, 1782.

IT is univerfally acknowledged, that the firft fimple principles of fcience cannot be too critically examined, in order to their being firmly eftablifhed; more efpecially thofe which relate to the practical and operative parts of mechanics, upon which much of the active bufinefs of mankind depends. A fentiment of this kind occafioned the preceding tract upon *Mechanic Power*. What I have now to offer was intended as a fupplement thereto, and the experiments were, in part, tried; but the completion thereof was deferred at the time of its firft publication, partly from want of leifure; partly to avoid too great a length of the paper itfelf; and partly to avoid the bringing forward too many points at once.

My prefent purpofe is to fhew, that the true doctrine of the *collision of bodies* hangs as it were upon the fame hook, as the doctrine of the gradual generation of motion from reft, confidered in that paper: that is, that whether bodies are put into gradual motion, and uniformly accelerated from reft to any given velocity; or are put in motion, in an inftantaneous manner, when bodies of any kind ftrike one another; the motion, or fum of the motions produced, has the fame relation to mechanic power therein defined, which is neceffary to produce the motion defired. To prove this, and at the fame time to fhew fome capital miftakes in principle, which have been *assumed* as indifputable truths by men of great learning, is the reafon of my now purfuing the fame fubject.

I do not mean to point out the particular miftakes which have been made by particular men, as that would lead me into too great a length: I fhall therefore content myfelf with obferving, that the laws of collifion, which have been inveftigated by mathematical philofophers, are principally of three kinds; viz. thofe relating to bodies perfectly *elastic*; to bodies perfectly unelaftic, and perfectly *soft*; and to bodies perfectly unelaftic, and perfectly *hard*. To avoid prolixity, I fhall confider in each, only the fimple cafe of two bodies which are equal in weight or quantity of matter ftriking one another. Refpecting thofe which are perfectly elaftic, it is univerfally agreed, that when two fuch bodies ftrike one another, no motion is loft; but that in all cafes, what is loft by by one is acquired by the other: and hence, that if an elaſtic body in motion ſtrikes another at reſt, upon the ſtroke the former will be reduced to a ſtate of reſt, and the latter will fly off with an equal velocity.

In like manner, if a non-elaſtic *ſoft* body ſtrikes another at reſt, they neither of them remain at reſt, but proceed together from the point of colliſion with exactly one half of the velocity that the firſt had beroie the ſtroke ; this is alſo univerſally allowed to be true, and is fully proved by every good experiment upon the ſubject.

Reſpecting the third ſpecies of body, that is, thoſe that are non elaſtic and yet perfectly hard : the laws of motion relating to them, as laid down by one ſpecies of philoſophers, have been rejected by another; the latter alledging, that there are no ſuch bodies to be found in nature whereon to try the experiment; but thoſe who have laid down and aſſigned the doctrine that would attend the colliſion of bodies, of this kind (if they could be found) have univerſally agreed, that if a non-elaſtic *hard* body was to ſtrike another of the ſame kind at reſt, that, in the ſame manner as is agreed concerning non-elaſtic ſoft bodies, they neither of them would remain at reſt, but would in like manner proceed from the point of colliſion, with exactly one half of the velocity that the firſt had before the ſtroke : in ſhort, they lay it down as a rule attending all non-elaſtic bodies, whether hard or ſoft, that the velocity after the ſtroke will be the ſame in both, viz. *one half* of the velocity of the original ſtriking body.

Here is therefore the aſſumption of a principle, which in reality is proved by no experiment, nor by any fair deduction of reaſon that I know of, viz. that the velocity of non-elaſtic *hard* bodies after the ſtroke muſt be the ſame as that reſulting from the ſtroke of non-elaſtic *ſoft* bodies ; and the queſtion now is, whether it be true or not?

Here it may be very properly aſked, what ill effects can reſult to practical men, if philoſophers ſhould reaſon wrong concerning the effects of what does not exiſt in nature, ſince the practical men can have no ſuch materials to work upon, or miſ-judge of? But it is anſwered, that they who infer an equality of effects between the two ſorts, may from thence be miſled themſelves, and in conſequence miſlead practical men in their reaſonings and concluſions concerning the ſort with which they have abundant concern, to wit, the non-elaſtic *ſoft bodies,* of which water is one, which they have much to do with in their daily practice.

Previous

Previous to the trying my experiment on mills, I never had doubted the truth of the doctrine, that the fame velocity refulted from the ftroke of both forts of non-elaftic bodies; but the trial of thofe experiments made me clearly fee, at leaft the inconclufivenefs, if not the falfity of that doctrine: becaufe I found a refult which I did not expect to have arifen from either fort; and for the which, when it appeared from experiment, I could fee a fubftantial reafon why it fhould take place in one fort, and that it was impoffible that it could take place in the other; for if it did, the bodies could not have been perfectly *hard*, which would be contrary to the hypothefis: of this deduction I have given notice in my faid tract on mills, page 49. *The effect therefore of overshot wheels, &c.*

It may alfo be faid, that fince we have no bodies perfectly elaftic, or perfectly unelaftic and *soft*, why fhould we expect bodies perfectly unelaftic and *hard?* Why may not the effects be fuch as fhould refult from a fuppofition of their being *imperfectly elastic* joined with their being *imperfectly hard?* But here I muft obferve, that the fuppofition appears to be a contradiction in terms.

We have bodies which are fo nearly perfectly elaftic, that the laws may be very well deduced and confirmed by them; and the fame obtains with refpect to non-elaftic *soft* bodies; but concerning bodies of a mixed nature, which are by far the greateft number, fo far as they are wanting in elafticity, they *are soft* and *bruise, yield,* or *leave a mark* in collifion; and fo far as they are not perfectly foft they are elaftic, and obferve a mixture of the law relative to each; but imperfectly elaftic bodies, imperfectly hard, come in reality under the *same description* as the former mixed bodies: for fo far as they are imperfectly hard they are foft, and either *bruise* and *yield*, or leave a mark in the ftroke; and fo far as they want perfect elafticity, they are non-elaftic; that is to fay, they are bodies imperfectly elaftic, and imperfectly foft; and in fact I have never yet feen any bodies but what come under this defcription. It feems, therefore, that refpecting the *hardness* of bodies they differ in degrees of it, in proportion as they have a greater degree of tenacity or cohefion; that is, are further removed from perfect foftnefs, at the fame time that their elaftic fprings, fo far as they reach, are very ftiff; and hence we may (by the way) conclude, that the fame mechanic power that is required to change the figure in a *small degree* of thofe bodies that have the popular appellation of *hard bodies*, would change it in a *great degree* in thofe bodies that approach towards foftnefs, by having a fmall degree of tenacity or cohefion. In the former kind we may rank the harder kinds of *cast iron,* and in the latter, *soft tempered clay.*

O

While

While the philofophical world was divided by the difpute about the *old* and *new opinion*, as it was called, concerning the powers of bodies in motion, in proportion to their different velocities; thofe who held the old opinion contending, that it was as the velocity *simply*, afked thofe of the new, how, upon their principles, they would get rid of the conclufions arifing from the doctrine of unelaftic perfectly hard bodies? They replied, they found no fuch bodies in nature, and therefore did not concern themfelves about them. On the other hand, thofe of the new opinion afked thofe of the old, how they would account for the cafe of non-elaftic foft bodies, where, according to them, the whole motion loft by the ftriking body was retained in the two after the ftroke (the two bodies moving together with the half velocity), though the two non-elaftic bodies had been bruifed and changed their figure by the ftroke; for, if no motion was loft, the change of figure muft be an effect without a caufe? To obviate this, thofe of the old opinion ferioufly fet about proving, that the bodies might change their figure, without any lofs of motion in either of the ftriking bodies.

Neither of thefe anfwers have appeared to me fatisfactory, efpecially fince my mill experiments: for, with refpect to the firft, it is no proper argument to urge the impoffibility of finding the proper material for an experiment, in anfwer to a conclufion drawn from an abftract idea. On the other hand, if it can be fhewn, that the figure of a body can be changed, without a *power*, then, by the fame law, we might be able to make a *forge hammer* work upon a mafs of foft iron, without any other power than that neceffary to overcome the friction, refiftance, and original *vis inertiæ*, of the parts of the machine to be put in motion: for, as no progreffive motion is given the mafs of iron by the hammer (it being fupported by the anvil), no power can be expended that way; and if none is loft to the hammer from changing the figure of the iron, which is the only effect produced, then the whole power muft refide in the hammer, and it would jump back again to the place from which it fell, juft in the fame manner as if it fell upon a body perfectly elaftic, upon which, if it did fall, the cafe would really happen: the power, therefore, to work the hammer would be the fame, whether it fell upon an elaftic or non-elaftic body; an idea fo very contrary to all experience, and even apprehenfion, of both the philofopher and vulgar artift, that I fhall here leave it to its own condemnation.

As nothing, however, is fo convincing to the mind as experiments obvious to the fenfes, I was very defirous of contriving an experiment in point, and as I faw no hopes of finding matter to make a *direct* experiment, I turned my mind towards an indirect
one:

one : fo circumfcribed, however, as to prove inconteftably, that the refult of the ftroke of two non-elaftic perfectly hard bodies could not be the fame as would refult from the collifion of two foft ones ; that is, if it can be *bonâ fide* proved, that *one half* of the original power is loft in the ftroke of foft bodies by the change of figure (as was very ftrongly fuggefted by the mill experiments) ; then fince no fuch lofs can happen in the collifion of bodies perfectly hard, the refult and confequence of fuch a ftroke muft be *different.*

The confequence of a ftroke of bodies perfectly hard, but void of elaficity, muft doubtlefs be different from that of bodies perfectly *elastic :* for having no fpring the body at reft could not be driven off with the velocity of the ftriking body, for that is the confequence of the action of the fpring or elaftic parts between them, as will be fhewn in the refult of the experiments ; the ftriking body will therefore not be ftopped, and as the motion it lofes muft be communicated to the other, from the equality of action and re-action, they will both proceed together, with an equal velocity, as in the cafe of non-elaftic foft bodies : the queftion, therefore, that remains is, what that *velocity must be ?* It muft be greater than that of the non-elaftic foft bodies, becaufe there is no mechanical power loft in the ftroke. It muft be lefs than that of the ftriking body, becaufe, if equal, inftead of a *loss* of motion by the collifion it will be doubled. If, therefore, non-elaftic foft bodies lofe half their motion, or mechanical power, by change of figure in collifion, and yet proceed together with half the velocity, and the non-elaftic hard bodies can lofe *none* in any manner whatever ; then, as they muft move together, their velocity muft be fuch as to preferve the equality of the mechanic power, *unimpaired,* after the ftroke the fame as it was before it.

For example, let the velocity of the ftriking body before the ftroke be 20, and its mafs or quantity of matter 8 ; then, according to the rule deduced from the experiments in the tract on *Mechanic Power* (fee exp. 3d and 4th) that power will be expreffed by $20 \times 20 = 400$, which $\times 8 = 3200$; and if half of it is loft in the ftroke, in the cafe of non elaftic foft bodies, it will be reduced to 1600 ; which $\div$ 16 the double quantity of matter, will give 100 for the fquare of their velocity ; the fquare root of which being 10, will be the velocity of the two non-elaftic foft bodies after the ftroke, being juft one half of the original velocity, as it is conftantly found to be. But in the non-elaftic hard bodies, no power being loft in the ftroke, the mechanic power will remain after it, as before it, $= 3200$; this, in like manner, being divided by 16, the double quantity of matter, will give 200 for the fquare of the velocity, the fquare root

of

of which is 14.14, &c. for their velocity after the ftroke, which is to 10, the velocity of the non elaftic foft bodies after the ftroke, as the fquare root of 2 to 1, or as the diagonal of a fquare to its fide.

It remains, therefore, now to be proved, that precifely half of the mechanic power is *lost* in the collifion of non-elaftic foft bodies; for which purpofe my mind fuggefted the following reflections. In the collifion of elaftic bodies the effect feemingly inftantaneous, is yet performed in *time;* during which time the natural fprings refiding in elaftic bodies, and which conftitute them fuch, are bent or forced, till the motion of the ftriking body is divided between itfelf and the body at reft; and in this ftate the two bodies would then proceed together, as in the cafe of non-elaftic foft bodies; but as the fprings will immediately reftore themfelves in an equal time, and with the fame degree of *impulsive force,* wherewith they were bent in this re-action, the motion that remained in the ftriking body will be totally deftroyed, and the total exertion of the two fprings, communicated to the original refting body, will caufe it to fly off with the fame velocity wherewith it was ftruck.

Upon this idea, if we could conftruct a couple of bodies in fuch a way that they fhould either act as bodies perfectly elaftic; or that their fprings fhould at pleafure be hooked up, retained, or prevented from reftoring themfelves, when at their extreme degree of bending; and if the bodies under thefe circumftances obferved the laws of collifion of non-elaftic foft bodies, then it would be proved, that one half of the mechanical power, refiding in the ftriking body, would be loft in the action of collifion; becaufe the impulfive force or power of the fpring in its reftitution being cut off, or fufpended from acting, which is equal to the impulfive force or power to bend it (and which alone has been employed to communicate motion from one body to the other), it would make it evident, that one half of the impulfive force is loft in the action, as the other half remains *locked up* in the fprings, it alfo follows, as a *collateral circumstance,* that be the impulfive power of the fprings what it may from firft to laft, yet as one half of the *time* of the action is by this means cut off, in this fenfe alfo it will follow, that one half of the mechanic power is deftroyed; or rather, in this cafe, remains locked up in the fprings, capable of being *re-exerted* whenever they are fet at liberty, and of producing a frefh mechanical effect, equivalent to the motion or mechanical power of the two non-elaftic foft bodies after their collifion.

Hence we muft infer, that the quantity of mechanical power expended in difplacing the parts of non-elaftic foft bodies in collifion, is exactly the fame as that expended
in

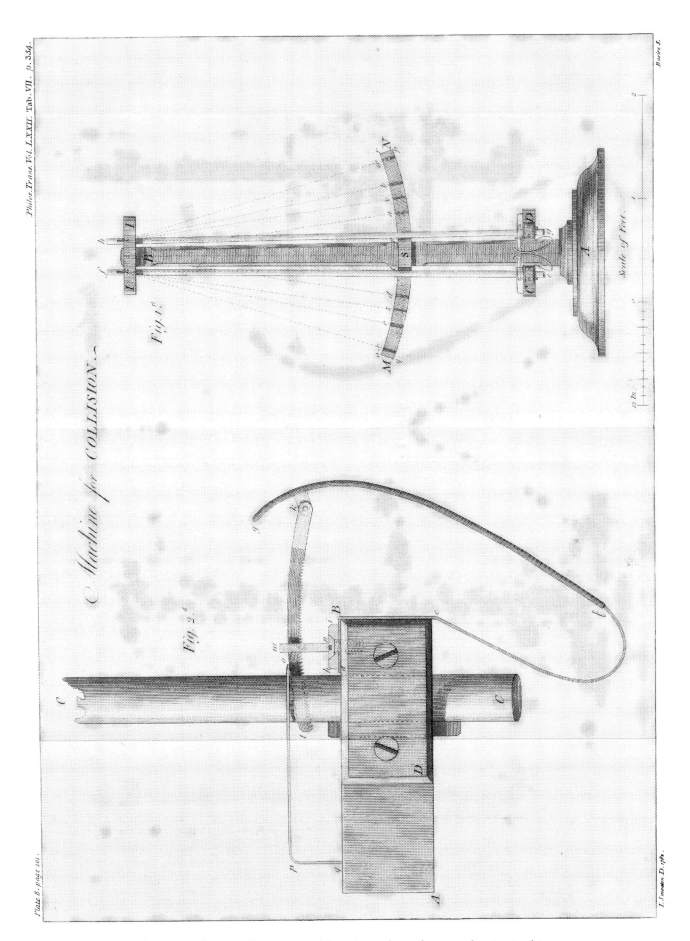

Machine for COLLISION.

Fig. 1.ᵗ

Fig. 2.ᵈ

Scale of Feet.

Plate 8. page 101.

The material originally positioned here is too large for reproduction in this reissue. A PDF can be downloaded from the web address given on page iv of this book, by clicking on 'Resources Available'.

in bending the fprings of perfectly elaftic bodies; but the difference in the ultimate effect is, that in the non-elaftic foft bodies, the power taken to difplace the parts will be totally loft and deftroyed, as it would require an equal mechanic power to be raifed a-frefh, and exerted in a contrary direction to reftore the parts back again to their former places; whereas, in the cafe of the elaftic bodies, the operation of half the mechanic power is, as obferved already, only locked up and fufpended, and capable of being re-exerted without a further original acceffion.

Thefe ideas arofe from the refult of the experiments tried upon the machine defcribed in my faid tract upon mechanic power, and were alfo communicated to my very worthy and ingenious friend WILLIAM RUSSEL, Efq. F. R. S. at the fame time that I fhewed him thofe experiments in 1759; but the mode of putting this matter to a full and fair mechanical trial has fince occurred; and though fome rough trials, fufficient to fhew the effect, were made thereon, prior to the offering the Paper on Mechanical Power to the Society in 1776, yet the machine itfelf I had not leifure to complete to my fatisfaction till lately; which I mention to apologize for the length of time that thefe fpeculations have taken in bringing forward.

Description of the Machine for Collision.

PLATE VIII. Fig. 1. fhews the front of the machine as it appears at reft when fitted for ufe.

A is the pedeftal, and A B the pillar, which fupports the whole, C D are two compound bodies of about a pound weight each, but as nearly equal in weight as may be. Thefe bodies are alike in conftruction, which will be more particularly explained by fig. 2. Thefe bodies are fufpended by two white fir rods of about half an inch diameter, *e f* and *g h* being about four feet long from the point of fufpenfion to the centre of the bodies; and their fufpenfion is upon the crofs piece I I, which is mortifed through, to let the rods pafs with perfect freedom; and they hang upon two fmall plates filed to an edge on the under fide, and pafs through the upper part of the rods. Their centres are at *k* and *l*, and the edges being let into a little notch, on each fide the mortife, the rods are at liberty to vibrate freely upon their refpective points (or rather edges) of fufpenfion, and are determined to one plane of vibration. M N is a flat arch of white wood, which may be covered with paper, that the marks thereupon may be more confpicuous.

The

The crofs piece I I is made to project fo far before the pillar, that the bodies in their vibrations may pafs clear of it, without danger of ftriking it; and alfo the arch M N is brought fo far forward as to leave no more than a clearance, fufficient for the rods to vibrate freely without touching it.

Fig. 2, fhews one of the compound bodies, drawn of its full fize. A B is a block of wood, and about as much in breadth as it is reprefented in height, through a hole in which the wood rod C C paffes, and is fixed therein.

D B reprefents a plate of lead about three-eighths of an inch thick, one on each fide, fcrewed on by way of giving it a competent weight, d B $e\,f\;g$ reprefents the edge of a fpringing plate of brafs, rendered elaftic by hard hammering; it is about five-eighths of an inch in breadth, and about one-twentieth of an inch thick. It is fixed down upon the wooden block at its end d B by means of a bridge plate, whofe end is fhewn $h\,i$, and is fcrewed down on each fide of the fpring plate by fcrews, which being relaxed the fpring can be taken out at pleafure, and adjufted to its proper fituation. $k\,l$ is a light thin flip of a plate, whofe under edge is cut into teeth like a fine faw or ratchet, and is attached to the fpring by a pin at k, which paffes through it, and alfo through a fmall ftud rivetted into the back part of the fpring, and upon which pin, as a centre, it is freely moveable. $m\,n$ fhews a fmall plate or ftud feen edgeways raifed upon the bridge plate, through an hole in which ftud the ratchet paffes; and the lower part of the hole is cut to a tooth fhaped properly to catch the teeth of the ratchet, and retain it together with the fpring at any degree to which it may be fuddenly bent; and for this intent it is kept bearing gently downward, by means of a wire fpring $o\,p\,q$, which is in reality double, the bearing part at o being femi-circular; from which branching off on each the rod C C, paffes to p, and fixes at each end into the wood at q. However, to clear the ratchet, which is neceffarily in the middle as well as the rod, the latter is perforated; and alfo the block is cut away, fo far as to fet the mainfpring at e free of all obftacles that would prevent its play from the point B. The part $f\,g$ is fhewn thicker than the reft, by being covered with thin kid leather tight fewed on, to prevent a certain jarring that otherwife takes place on the meeting of the fprings in collifion.

Let us now return to fig. 1, the marks upon the arch M N are put on as follows. $o\,p$ is an arch of a circle from the centre l, and $q\,r$ an arch of a circle from the centre k, interfecting each other at S. Now the middle line of the marks t, v, are at the fame diftance from the middle line at S that the centres $k\,l$ are; fo that when

each

each body hangs in its own free pofition, without bearing againft the other, the rod *e f* will cover the mark at *t*, and the rod *g h* will cover the mark at *v*. From the point S upon the arches S *p* and S *q* refpectively, fet off points at an equal and competent diftance from S each way, which will give the middle of the mark *w* and *x*: and upon the arch S *p* find the middle point between the mark *v* and *w*, which let be *y*; and on the other fide, in like manner, upon the arch S *q* find a middle point for the mark *z*; then fet off the diftance S *v* or S *t* from *y* each way, and from *z* each way; and from thefe points, drawing lines to the refpective centres *l* and *k*, they will give the place and pofition of the marks *a*, *b*, and *c*, *d*; and thus is the machine prepared for ufe.

For Trials on elastic Bodies:

For this ufe take out the pins and ratchets from each refpectively, and the fprings being then at liberty, with a fhort bit of ftick (fuppofe the fame fize as the rods) turn afide the rod *g h* with the right hand, carrying the body D upwards till the ftick is upon the mark *w*, as fuppofe at o; there hold it, and with the left fet the body C perfectly at reft; in which cafe the rod *e f* will be over the mark *t*; then fuddenly withdraw the ftick, in the direction that the rod *g h* is to follow it, and the fpring of the body D, impigning upon that of the body C, they will be both bent, and alfo reftored; and the body C will fly off, and mount till its rod *e f* covers the mark *x*; the rod of the ftriking body D remaining at reft upon its proper mark of reft *v*, till the body C returns, when the body D will fly off in the fame manner; the two bodies thus rebounding a number of times, lofing a part of their vibration each time; but fo nearly is the theory of elaftic bodies fulfilled hereby, that the fingle advantage of originally pufhing the rod *g h* beyond the mark *w*, by the thicknefs of the ftick, or its own thicknefs, is fufficient to carry the rod of the quiefcent body C completely to its mark *x*.

There are feveral other experiments which may be made with this apparatus, in confirmation of the doctrine of the collifion of elaftic bodies; which being univerfally agreed upon, and well known, it is needlefs further to dwell upon here; but refpecting the application to non-elaftic foft bodies, it is far more difficult to come at a fitnefs of materials for this kind of experiment, than it is for thofe fuppofed perfectly elafticity. The conclufions, however, may be attained with equal certainty.

For

For Trials on Non-elastic soft Bodies :

For this purpofe the ratchets muft be applied and put in order as before defcribed, and the fprings being both put to their point of reft, let the body D be put to its mark *w* in the fame manner as before defcribed, and the body C to reft. The body D being let go, and ftriking the body C at reft, in confequence of the ftroke, the fprings being hooked up by the ratchets, they both move from their refting marks *t*, *v*, refpectively towards M : Now if they both moved together, and the rod *e f* covered the mark *c*, and the rod *g h* covered the mark *d* at their utmoft limit, then they would truly obey the laws of non-elaftic foft bodies ; becaufe their medium afcent would be to the mark *z*, which is juft half the angle of afcent to the mark *x* ; but as in this piece of machinery, though the main or principal fprings are hooked up, yet every part of them, and all the materials of which they are compofed, and to which they are attached, have a degree, or more properly fpeaking, a certain compafs of *elasticity*, which, as fuch, is perfect, and no motion is loft thereby.

We muft not, therefore, expect the two compound bodies after the ftroke to ftick together without feparating, as would be the cafe with bodies truly non-elaftic and foft ; but that from the elafticity they are poffeffed of, they will by rebounding be feparated ; but that elafticity being perfect, can occafion no lofs of motion to the fum of the two bodies ; fo that if the body C afcends as much above its mark *c* as the body D falls fhort of its mark *d*, then it will follow, that their medium afcent will ftill be to the mark *z*, as it ought to have been, had they been truly non-elaftic foft bodies ; and this, in reality, is truly the cafe in the experiment, as nearly as it can be difcerned.

After a few vibrations, by the rubbing of the fprings againft one another, they are foon brought to reft ; and here they would *always rest* had they been truly and pro-perly perfect non-elaftic foft bodies ; but here, as in the cafe of thefe bodies, by a change of the figure and fituation of the component parts, there is expended one half of the mechanical power of the firft mover, yet in this cafe the other half is not *lost*, but *suspended* ready to be re-exerted whenever it is fet at liberty ; and that it is really and *bonâ fide one half*, and neither more nor lefs appears from this uncontroverted fimple principle, that the power of reftitution of a perfect fpring is exactly equal to the power that bends it.   And this may, in a certain degree, be fhewn to be fact by experiment, if there were any need of fuch a proof ; for if, when the bodies are at reft after the laft experiment, the two rods are lafhed together at the bottom with a bit of thread, and then

the

the ratchets unpinned and removed; on cutting the thread with a pair of fciffars they will each of them rebound, C towards M and D towards N, and if they rebounded refpec- tively to *z* and *y,* the mechanical power exerted would be the fame as it was after the ftroke, when the mean of their two afcents was up to the mark Z; but here it is not to be expected, becaufe not only the motion loft by the friction of the ratchets is to be deducted, becaufe it had the effect of real non-elafticity; but alfo the elafticity that feparated them in the ftroke, which was loft in the vibrations that fucceeded; neither of which hindered the mean afcent to be to *z*; but yet, under all thefe difadvantages in the machine (if not unreafonably ill made) the rod *e f* will afcend to *d,* and *g h* to *a*: and hence I infer, as a pofitive truth, that in the collifion of non-elaftic foft bodies, *one half of the mechanic power refiding in the ftriking body is loft in the ftroke.*

Refpecting bodies unelaftic and perfectly hard, we muft infer, that fince we are un- avoidably led to a conclufion concerning them, which contradicts what is efteemed a truth capable of the ftricteft demonftration; viz. that the velocity of the centre of gravity of no fyftem of bodies can be changed by any collifion betwixt one another, fomething muft be affumed that involves a contradiction. This perfectly holds, ac- cording to all the eftablifhed rules, both of perfectly elaftic and perfectly non-elaftic *foft* bodies; rules which muft fail in the perfectly non-elaftic *hard* bodies, if their velocity after the ftroke is to the velocity of the ftriking body as one is to the fquare root of 2; for then the centre of gravity of the two bodies will by the ftroke acquire a velocity greater than the centre of gravity the two bodies had before the ftroke in that proportion, which is proved thus.

At the outfet of the ftriking body, the centre of gravity of the two bodies in our cafe will be exactly in the middle between the two; and when they meet it will have moved from their half diftance to their point of contact, fo the velocity of the centre of gravity before the bodies meet will be exactly one half of the velocity of the ftriking body; and, therefore, if the velocity of the ftriking body is 2, the velocity of the centre of gravity of both will be one. After the ftroke, as both bodies are fuppofed to move in contact, the velocity of the centre of gravity will be the fame as that of the bodies; and as their velocity is proved to be the fquare root of 2, the velocity of their centre of gravity will be increafed from 1. to the fquare root of 2; that is, from 1. to 1.414, &c.

The fair inference from thefe contradictory conclufions therefore is, that an unelaftic hard body (perfectly fo) is a repugnant idea, and contains in itfelf a contradiction; for,

P   to to make it agree with the fair conclusions that may be drawn on each side, from clear premises, we shall be obliged to define its properties thus: that in the stroke of unelastic hard bodies they cannot *possibly lose any* mechanic power in the stroke; because no other impression is made than the communication of motion; and yet they *must lose* a *quantity* of mechanic power in the stroke; because, if they do not, their common centre of gravity, as above shewn, will acquire an *increase* of velocity by their stroke upon each other.

In a like manner the idea of a *perpetual motion* perhaps, at first sight, may not appear to involve a contradiction in terms; but we shall be obliged to confess that it does, when, on examining its requisites for execution, we find we shall want bodies having the following properties; that when they are made to *ascend* against gravitation their absolute weight shall be *less*; and when they *descend* by gravitation (through an equal space) their absolute weight shall be greater; which, according to all we know of nature, is a *repugnant* or *contradictory idea.*

A DESCRIPTION

A DESCRIPTION of a new Tackle or Combination of Pullies, by Mr. JOHN SMEATON.

Philosophical Transactions, VOL. XLVII.

Read June 11, 1752. THE axis *in peritrochio,* and the tackle of pullies, are the only mechanic powers, which can, with convenience, be applied to the moving of large weights, when the height, to which they are intended to be raifed, is confiderable. The excellence of the former is, its working with little friction ; that of the latter, its being eafily moved from place to place, and applied *ex tempore,* as occafion requires; but when the weights are very large they are ufed in conjunction.

The prefent methods of arranging pullies in their blocks may be reduced to two * The firft confifts in placing them one by the fide of another, upon the fame pin ; the other in placing them directly under one another, upon feparate pins. But in each of thefe methods an inconvenience arifes, if above three pullies are framed in one block. For, according to the firft method, if above three pullies are placed by the fide of one another, as the laft part of the rope by which the draught is made (or, as it is commonly called, the fall of the tackle) muft neceffarily be upon the outfide pulley or fhieve ; the difference of the friction of the pulley will produce fo great a tendency to pull the block awry, that as much will be loft by the rubbing of the fhieves againft the block, on account of its obliquity, as will be got by increafing the number of pullies.

The fecond method is free from this objection ; but, as the length of the two blocks, taken together, muft be equal to the fum of the diameters of the pullies, befides the fpaces between for the ropes, and the neceffary appendages of the framing ; if there are more than three pullies in each block, they will run out into fuch an inconvenient length, as to deduct very confiderably from the height, to which the weight might otherwife have been raifed : fo that, upon thofe accounts, no very great purchafe can be made by the common fimple tackles of pullies alone.

---

* This is to be underftood of fimple tackles or purchafes by pullies, where one rope line or fall is reeved round and common to all.

In order, therefore, to increaſe their power on board a ſhip, a purchaſe is commonly made uſe of for loading and unloading of goods, called a *runner and tackle* : by this method the power of any ſimple tackle is doubled by the runner, and is very commodious there, as the maſts afford a ſufficient height; but as the runner block, upon which the whole weight depends, muſt be fixed at more than double the height to which the body is intended to be raiſed it is ſeldom uſed aſhore.

Sometimes a ſecond tackle is fixed upon the fall of the firſt; but here it is obvious, that whatever be the power of the ſecond tackle, the height to which the weight might otherwiſe have been raiſed by the firſt, will be leſs in the ſame proportion as the purchaſe is increaſed by the ſecond; unleſs it is ſhifted, which loſes time.

Again, very frequently the fall of the firſt tackle is applied to a capſtan or windlaſs, which increaſes the purchaſe very commodiouſly without the inconveniencies laſt-mentioned; but then the machine is rendered more cumberſome, and, conſequently, leſs fit for a moveable apparatus.

Thoſe impediments I have avoided, by combining the two methods of managing pullies firſt deſcribed. The pullies in this new combination are placed in each block in two tiers; ſeveral being upon the ſame pin, as in the firſt method, and every one having another under it, as in the ſecond; as alſo that, when the tackle is in uſe, the two tiers, that are the remoteſt from one another, are ſo much larger in diameter than thoſe that are neareſt, as to allow the lines of the former to go over the lines of the latter without rubbing.

From this conſtruction ariſes a new method of reeving the line: for here, let the number of ſhieves be what it will, the fall of the tackle will always be upon the middle ſhieve, or on that next the middle, according as the number of pullies on each pin is odd or even.

To do this, the line is fixed to ſome convenient part of the upper block, and brought round the middle ſhieve of the larger tier of the under block, or the ſhieve neareſt to the middle if the number is even, from thence round one of the ſame ſort, *next* to the centre one, of the upper block; and ſo on till the line comes to the outſide ſhieve, where the laſt line of the larger tier falls upon the firſt ſhieve of the ſmaller, and being reeved round thoſe, till it comes to the oppoſite ſide; the line from the laſt ſhieve of the ſmaller tier, again riſes to the firſt of the larger, whence it is conducted round the remainder of the larger tiers till it ends on the middle fhieve of the upper block; on the larger tier, or the fhieve neareft to the middle, if the number is even, as will appear more plain, by infpection of figure 6, PLATE I.

In this method all the lines or parts are clear of one another, and the friction being more equally divided, the blocks are kept parallel.

The model in brafs which I have the honor to fhew the Society, and from which I made the draught, confifts of 20 fhieves, five on each pin. With this model, which may eafily be carried in the pocket, I have raifed fix hundred weight. But with a tackle of this fort, properly executed in large, one man will eafily raife a ton, and a greater number of men a greater weight, in proportion*.

I have tried feveral numbers of fhieves as far as 36; but 20 feems to be the largeft number that will anfwer well in practice.

By this tackle all moderate weights (that is, from fix hundred weight to two or three tons) can be raifed to any moderate height, without being compounded with any other purchafe, and without any lofs of height more than the length of the two blocks, which is common to all purchafes with pullies whatever; and without requiring more room than the fpace neceffary for the afcent of the body: this tackle is eafily moved from place to place, and where neceffary, can be applied with equal advantage in compound purchafes, in the fame manner as any other fimple tackle.

Large weights can alfo be raifed by bringing the fall of the tackle to a capftan or windlafs, and weights three times heavier than can be raifed in the common way, may be raifed by this method, becaufe the fize of the largeft rope being given, that can be ufed as a tackle fall, 20 of thofe ropes will bear more than three times the weight that fix will bear; and laftly, the weight being given, the tackle fall may be lefs in the proportion of 20 to 6, which will render the whole more flexible, and as the block pins can be confiderably lefs, the friction will be lefs alfo, and the tackle more eafily over hauled.

---

* A large tackle of 20 was tried on board one of his majesty's ships; and by the help thereof, though it was with a new rope, one man raised one of the ship-guns and carriage, that together weighed 27 hundred weight; there being a person, as usual, to *hold on*, or prevent the rope from slipping back..

A very

A very commodious tackle of 12 may be executed in wood, in the fame manner that common blocks are made *.

I fhould not have troubled the Royal Society with an account of this contrivance, did it not feem promifing of fome utility, in a variety of purpofes; particularly for merchants, feamen, builders, engineers, &c. I therefore entirely fubmit it to the cenfure of that honorable body.

J. SMEATON.

P. S. In conftructing in wood a tackle of 20 for 3 tons, the larger tier of fhieves fhould not be lefs than 10 inches; the running line needs not be more than 2 inches in circumference, and the pins, if iron, not exceeding five-eighths, if ebony, three-fourths of an inch in thicknefs.

---

* All the principal purchases used in building *Edystone Light-house*, with *stone*, are of this kind, and which are subject to no material inconvenience, but, on the contrary, for wet and dry, and rough usage, wooden pins, and holes both turned true and smooth, seem preferable to Mr. HALL, the same hand as also executed a large tackle of 20 in wood with success.

A DISCOURSE

A DISCOURSE concerning the Menstrual Parallax, arising from the mutual Gravitation of the Earth and Moon, its Influence on the Observation of the Sun and Planets, with a Method of observing it, by J. SMEATON, F. R. S.

Read May 12, 1768. IT is demonstrated by Sir ISAAC NEWTON, in the *Principia*, that it is not the earth's centre, but the common centre of gravity of the earth and moon, that describes the ecliptic; and that the earth and moon revolve in similar ellipses, about their common centre of gravity.

The same great author has also investigated, from the different rise of the tides, when the moon is in conjunction or opposite to the sun, to those which happen when the moon is in her quadratures; that the quantity of the matter in the earth is to that in the moon, as 39, 78 to 1; from whence, and the known distance of the earth and moon, it would follow, that the common centre of gravity of the two bodies falls without the surface of the earth, by one half of its semi-diameter, that is, that the centre of the earth describes an epicycle round the common centre of gravity once a month, whose diameter is three semi-diameters of the earth.

Dr. GREGORY, in his astronomy, has laid hold of this circumstance, in order to prove the relative gravity of the earth and moon, by observation; which is the subject of his sixtieth proposition of the fourth book; in which he has demonstrated, that if an observer on the earth makes a correct observation on the sun's place, when the moon is in one quadrature, it will differ from a like observation, taken in the opposite quadrature (according to a mean elliptic motion) by an angle which the diameter of this epicycle will subtend at the sun. The same learned author has also shewn, in the scholium to the same proposition, that this quantity, or parallax, will be twice greater to Mars in opposition, and three times greater to Venus, in her inferior conjunction with the sun.

The difference thus produced in the apparent place of the sun, and of all the primary planets, being governed by the moon, and having its period the same, may perhaps not be unaptly called *menstrual parallax*.

Now, if, with Sir ISAAC NEWTON, the relative gravities of the earth and moon are taken between the proportion of 39 and 40 to one, the menstrual parallax of the sun will

will come out 13 upon the radius of the earth's epicycle, and will affect the folar obfervations at the oppofite quadratures, by double that quantity, viz. 26″ in like manner, the mean diftance from the earth of Mars in oppofition, being to the fun's mean diftance as 1 to 2.1; and the leaft diftance of Mars from the earth to the fun's mean diftance, as 1 to 2¾, the menftrual parallax of Mars will affect the obfervation upon him in that fituation, by 56″ and 73″½ refpectively.

The mean diftance of Venus from the earth, in her inferior conjunction, being to that of the fun as 3½ to 1 nearly, and not very variable, on account of the orbit of Venus being almoft circular; the menftrual parallax would affect the place of Venus in that fituation, by a quantity not lefs than 92″; and in all other fituations in proportion to her diftance; which alfo holds with refpect to all the reft of the planets.

Thefe difturbing quantities are by no means to be difpenfed with in the nice and critical ftate that aftronomical obfervations and calculations have arrived at, in confequence of the difcoveries of Dr. BRADLEY, who may be faid to have given a bafis to aftronomy; however, could we rely upon the *data*, on which Sir ISAAC's inveftigation of the relative gravity of the earth and moon is founded, we fhould have nothing to do but to apply an equation to the particular cafes, according to the diameter of the epicycle, as deduced from the relative gravity; but whoever confiders the great obftructions that the water of the fea meets with in its motion, to obey the influence of the moon; the great difficulty in afcertaining the true height of the tides, from the many difturbing caufes intervening; and the many uncertainties, and the want of coincidence that have attended, and muft attend fuch obfervations; muft confefs that this matter does not feem capable of fuch a determination from that quarter as the prefent ftate of aftronomy requires.

Accordingly, fince the time of Dr. GREGORY, thofe great aftronomers Dr. BRADLEY, DE LA CAILLE, and others, have applied themfelves to determine the quantity of the menftrual parallax, from folar obfervations; but though thefe have given caufe to fuppofe that the relative gravity of the earth and moon are not above $\frac{2}{3}$ of the quantity deduced from the tides; yet, as the obfervation of thefe fmall angles principally depends upon the obfervation of the fun's right afcenfion (which, depending on the meafure of time, is lefs capable of exact obfervation, than if depending on divided inftruments); the deductions thence drawn feem ftill wanting of that certainty which the fubject demands; and if to this we add, from a deduction of Mr. MASKELYNE, that the relative gravity of the earth and moon is as 76 to 1, derived from the effect that the moon

produces

produces in the nutation of the earth's axis; the relative gravity, and, confequently, the parallaxes thereon depending, will be reduced to one half of thofe refulting from Sir Isaac's determination.

It is true, that the quantity of effect of the menftrual parallaxes will not be great, if computed upon Mr. Maskelyne's induction, for as much as that the common centre of gravity will be confiderably within the earth's furface; yet, even in that cafe, the fun's tranfit over the meridian, when the moon is in one quadrature, will differ nearly one fecond of time from that obferved from the oppofite quadrature; and though De la Caille and Mayer have formed equations depending on the moon, to be applied to the equation of time; yet, if we are at an uncertainty, whether the *maximum* is one fecond, two thirds of a fecond, or half a fecond of time, each way, we are ftill under a material difficulty; for though thefe differences are fo fmall, that it is not eafy to determine them exactly from folar obfervations; yet, as they are capable of creating a fenfible difference in thefe obfervations, they will, fo long as they remain un-determined, prevent that folidity and firmnefs to the folar obfervations, which is the more neceffary as they are the foundation of all the reft; but with refpect to thofe planets, that in their periods come nearer to us than we to the fun, the obfervations upon them will be effected by a greater uncertainty.

The determination of the menftrual parallax is of ftill more importance, as it is a neceffary confideration in the determination of the fun's parallax; and this, whether deduced from Mars or Venus, as I fhall prefently fhew more particularly; but at firft I muft ftate the quantity of the menftrual parallax, according to the beft *data* yet known, by a contrary procefs; and, taking the mean quantity of the fun's parallax according to the determination of Mr. Short, at 8″ 8, and the relative gravities of the earth and moon, according to Mr. Maskelyne, as 76-to 1, and the mean diftance of their centres equal to $60\frac{1}{2}$ femi-diameters, we fhall then have the diftance of the earth's centre from the centre of gravity, at $\frac{8}{10}$ of the earth's femi-diameter, (that is, $\frac{1}{5}$ of that femi-diameter within the earth's furface), and the menftrual parallax equal to $\frac{8}{10}$ of the fun's parallax; confequently about 7″; and the double menftrual parallax, or vacillation, arifing from the whole diameter of the epicycle 14″; the mean menftrual parallax of Mars in oppofition 29″$\frac{1}{2}$; the greateft 38″$\frac{1}{2}$; and that of Venus 49″; from hence it follows, that, were a perfon to attempt the fun's parallax, by the diurnal motion of the earth, applied as a bafis to Mars in oppofition, as has been formerly tried; and fhould the moon be at new or full at the fame time, the change of place of the earth's centre, in its own epicycle, would amount to an angle feen from Mars of 1″ 3 nearly; that is,

Q

in

in cafe the interval between the obfervations were eight hours, and Mars at his mean diftance, this change would in the fame time amount to 1″ 7 nearly. In like manner, if a tranfit of Venus happen near the new or full moon (as will be the cafe next year), the time of the tranfit will be effected by a change of place, fuch as the earth's centre will defcribe in its epicycle during the time of the whole tranfit, if the beginning and end are obferved in the fame place, or during the difference of abfolute time, at which the tranfit appears to begin or end to different obfervers in diftant meridians. Thus. when the fame obferver fees the beginning and end in the fame place, the cafe defcribed by that obferver, from the earth's diurnal motion, muft be corrected by the fpace defcribed by the earth's centre, in the circumference of its epicycle, during that time; which, if it be fuppofed of feven hours, will amount to an angle 1″ 9, feen from Venus but, where the beginning or end is feen by different obfervers in diftant meridians, as the difference of abfolute time can hardly amount to above fifteen minutes, the change of place of the earth's centre will for that time be but fmall; however, at the rate before mentioned, it will for fifteen minutes affect the parallactic angle feen from Venus by about $\frac{7}{100}$ of a fecond; and the parallax of the fun, by about $\frac{1}{100}$ part of the whole: but this proportional part will remain the fame, whether the diftance of meridians be fuch as produce a greater or lefs difference of abfolute time than fifteen minutes *.

From what has been faid, I fuppofe it will appear, that the effects of the menftrual parallax are worthy of confideration, and that nothing has been yet executed, whereby it has received a determination fufficiently accurate; for, in regard to obfervations upon the fun, the whole quantity is too fmall to be minutely obferved in right afcenfion; and with refpect to the application to Mars and Venus, has not been as yet fo critically reduced to computation, as to render their parallaxes (though in themfelves much greater) deducible with equal certainty as that of the fun.

What I have now to propofe, is a method of obferving the menftrual parallaxes of Mars and Venus, without laying an undue ftrefs upon the theory of their motions.

The firft opportunity of making an obfervation for this purpofe, will be at the next oppofition of Mars, which, according to the nautical almanack, will happen the 26th of

---

* If an error of $\frac{1}{300}$ part of the whole may be fuppofed in the obfervation for determining the fun's parallax by the transit of Venus, a neglect of the menftrual parallax may make it $\frac{1}{100}$ part of the whole.

October

October next, in the morning; I will therefore endeavour to illuftrate this matter by taking that as an example.

The diftance of Mars from the earth will then be fomewhat lefs than the mean diftance, that is, as 1 to 2.2; and, confequently, his double menftrual parallax, according to Mr. Maskelyne, will be nearer 31″ in the point of oppofition. Now, as the moon will be at full not above twelve hours preceding that oppofition, the moon will be nearly in the moft favourable fituation for the purpofe.

For this end, let an accurate obfervation be made upon the place of Mars at the following times, viz. firft, near the time of the new moon, preceding Mars's oppofition, or more properly at the neareft opportunity, to the time of the moon's oppofition to Mars, which will happen in the night, between the 12th and 13th of October; fecondly, let the place of Mars be obferved when the moon is neareft her quartile with Mars; that is, between the 19th and 20th of the fame month; thirdly, let an obfervation be made on Mars when the moon is in conjunction with Mars, the neareft to his oppofition with the fun, that is, between the 25th and 26th of October; fourthly, let Mars be again obferved when the moon has moved on to her quartile with Mars, viz. between the 31ft of October and 1ft of November; and fifthly and laftly, let the place of Mars be obferved when the moon has again got to her oppofition with Mars, which happens between the 7th and 8th of November.

Now it is manifeft, that when the moon is in conjunction or oppofition to Mars, the centre of the earth, the centre of Mars, and the common centre of gravity of the earth and Mars, will be nearly in a right line, and, confequently, that an obferver will then fee Mars, in the fame place in the heavens, as if the common centre of gravity were the fame as the centre of the earth; therefore, the place of Mars will then be unaffected by a menftrual parallax; and fuch will be the firft, third, and fifth, of the obfervations above propounded.

It is equally evident, that when the moon is in quartile with Mars, and moving towards a conjunction, an obferver, at the earth's centre, will fee Mars more backward in the ecliptic, than if feen from the common centre of gravity, by $15''\frac{1}{2}$; and that, when the moon is in her oppofite quartile with Mars, and moving from her conjunction, an obferver at the earth's centre will fee Mars advanced in his orbit more forward by $15''\frac{1}{2}$, than if feen from the common centre of gravity; and the one obferva-

tion

tion checked with the other, will, according to a mean elliptic motion, differ by the quantity of 31″; and such will be the second and fourth observations above propounded.

Now, from the first, third, and fifth observations, three points of Mars' orbit will be given; which, by the help of the theory of Mars's motion in an elliptic orbit, whose aphelion, eccentricity, and nodes, are known sufficiently near for this purpose, the intermediate places of Mars may be inferred with the requisite degree of accuracy; and particularly, as the two intermediate observations, viz. the second and fourth, will be nearly at equal intervals of time between the three others: from hence it follows that the difference between the inferred or computed places, at the quartiles, and the observed places at those times will be the menstrual parallax required.

It is to be noted, that the times above specified are the most favourable for the observation; and could those be made uninterruptedly from weather, there would be the less occasion for any other; but as much as possible to prevent disappointment of this kind, it will be right to begin the observations a month preceding, making the proper observations, at the conjunctions, quartiles, and oppositions of the moon with Mars, which will be the means of supplying such observations as may happen to prove abortive before the opposition of Mars, and also, in case any of the observations to be made after that opposition shall prove deficient, the observations may be carried on a month or competent time afterwards.

As a further security against disappointments, as well as check, it will also be advisable to make proper observations the night preceding, and subsequent to those in which the quartiles, conjunction, &c. happen; for, as the quantities will not differ considerably from those obtained on the days specified, with proper allowances they may be brought in support and confirmation of the former.

In like manner, when Venus is moving towards her inferior conjunction with the sun, as will happen next year, the same observations may be made with respect to her, and continued for a necessary time to get observations of the place of Venus; viz. the first, when the moon is in conjunction or opposition with Venus: a second, when the moon is in her quartile with Venus; a third, in conjunction or opposition; a fourth, when the moon is in her opposite quartile to the former; and a fifth, again in conjunction or opposition:

poſition: the ſame opportunity will alſo offer when Venus is moving from her inferior conjunction with the ſun, and becomes a morning ſtar.

In regard to the obſervation of Venus, it is remarked by aſtronomers, that ſhe is to be ſeen with a good tranſit teleſcope, when ſhe is within a few degrees of the ſun; but as ſhe is three times nearer the earth, than the ſun's mean diſtance, when her elongation is 25° in the inferior part of her orbit, it is plain, that the neceſſary obſervations may be eaſily made when her menſtrual parallax will be at a medium, three times greater than the ſun's; and, conſequently, amounting for the whole difference 42″.

To avoid embarraſſment in deſcription, I have hitherto ſuppoſed that all the obſervations are made in the meridian; in which caſe the right aſcenſions will be the ſame as they would appear from the centre of the earth, and, conſequently, the planet's longitudes thence deduced nearly the ſame; but 'tis eaſy to ſee, that if the quartile obſervations are made when the planets are conſiderably to the eaſt or weſt of the meridian, and ſo choſen, that the place of the obſerver be further diſtant from the common centre of gravity, than the centre of the earth is from that centre, that the caſe of the obſervations will be conſiderably enlarged. Thus, in our latitude, ſuppoſing that the quartile obſervations are made four hours before and four hours after the planet paſſes the meridian, this will produce an enlargement of the baſis by one of the earth's ſemi-diameters; and as the whole baſe or diameter of the epicycle comes out, according to Mr. MasKELYNE, no more than 1. 6 of the earth's ſemi-diameters, the baſe will, according to this method, come out 2. 6; and, conſequently, at the next oppoſition, the menſtrual parallax of Mars will be thereby enlarged to 50″, the greateſt to 62″½, and that of Venus at a mean to 74″½.

It muſt however be acknowledged, that no kind of obſervations of the places of the planets are of equal validity with thoſe taken with the beſt inſtruments in the meridian; thoſe taken with micrometers perhaps not excepted; for, however accurately ſmall diſtances can be meaſured with the micrometer of Mr. Dolland, yet, as theſe meaſures can hardly be reduced to the ecliptic, without having the difference of declination or right aſcenſion from other means, (except two ſtars making ſomewhat near a right angle with the planet, ſhould appear within the field of view at once); and as in all theſe caſes the rectification of the places of the ſtars themſelves ultimately depends on meridian obſervations, we may perhaps be allowed to ſay, that in the moſt favorable caſes of the micrometer, the determinations thence to be drawn are

not

not fuperior to meridian obfervations, and in lefs favorable cafes muft be inferior; however, as the micrometer obfervations out of the meridian give an opportunity of repetition as often as we pleafe, and the obfervations for the rectification of the ftars concerned, can be repeated in the meridian as often as we pleafe alfo, it muft be equally allowed, that when thefe kind of obfervations are taken, not too near the horizon, when proper ftars offer for this purpofe, and the whole fkilfully managed, thefe kind of obfervations fall but little fhort of thofe taken immediately in the meridian.   I cannot therefore hefitate to recommend, that the quartile obfervations be taken out of the meridian, as well as in it in the firft place, by Dolland's micrometer, if ftars offer in proper pofitions, and if not, fecondly, by taking differences of right afcenfion and declination between the planet and the ftars, by the common micrometer, in cafe proper ftars offer themfelves for this purpofe; but as it frequently happens, that no proper ftars offer themfelves to micrometers of either kind, and this is ftill more likely to happen in the obfervations of Venus, which will be chiefly in the day light, I beg leave to offer (what to me is) a new method of obfervation out of the meridian; and which, though I efteem it not equal to micrometer obfervations of either kind, I apprehend will fall fo little fhort thereof, and prove fo much fuperior to any other method now in practice in thefe cafes, that I hope I fhall, on this occafion, be excufed in giving a particular defcription thereof; but, as it is a general method of obferving out of the meridian, I fhall referve it by way of appendix.

In the next obfervation of Mars, it has been ftated that, in the meridian obferva-tions alone, the menftrual parallax, according to the fmalleft eftimation, may be ex-pected to amount to 31″ in longitude, which, turned into right afcenfion, will make about 2″ of time : now, if it may be allowed, that a well practifed obferver can take the time of a tranfit to one quarter part of a fecond, over a fingle wire, if he has three wires, or more, as ufual, the mean of the three fhould be within $\frac{1}{12}$ part of a fecond; or within $\frac{1}{12}$ part of the whole quantity in queftion; it is, however, a matter of chance, whether the mean of three may or may not be within one-third part of the whole; and, as equal errors may be committed in the obfervations of the tranfits of the ftars, wherewith the right afcenfions of the planets in queftion are compared, which it is an equal chance, whether they tend to correct or increafe the errors committed in the former, yet, if, as has already been propofed, the obfervations are continued for two or three months, inftead of one, and obfervations, taken the preceding day and fub-fequent to the days of conjunction, quartile, and oppofition; and this, as well out of the meridian as in it, we can hardly doubt, but that, if the weather fhould favor, fo many checks would be formed, that, from the next oppofition of Mars alone, the

affair

affair may be brought within a twenty-fourth part of the whole; and, if to this be added the force of such determinations, as may be drawn from observations on Venus, before and after her transit over the sun next year, it can hardly be doubted, but that those three will bring us within a single second of a degree, subtended from the nearest planet; and these conclusions will be further strengthened by future observations, as two years will scarcely pass without affording one or more opportunities of this kind.

As I meant not to embarrass myself with exact computations, I have constantly supposed the distance of the common centre of gravity from the centre of the earth, to be a fixed quantity, whereas it will vary in the same proportion as the moon's distance varies; but as this and many other minutiæ will properly enter the computation, when the observations are made, I must beg leave to refer them to the learned in this science.

J. SMEATON.

Austhorpe, April 17,
1768.

DESCRIPTION

DESCRIPTION of a new Method of observing the Heavenly Bodies, out of the Meridian, by J. SMEATON, F. R. S.

Read May 16, 1768. THE inftrument I propofe for this purpofe, is a tranfit telefcope, mounted on a vertical axis; for example, fuch a one as is defcribed in the introduction to the Hiftoire Celefte, of Mr. Le Monnier, being one of the inftruments made by Mr. Graham, for the academicians, who went to meafure a degree at the polar circle; this, or any other inftrument upon equivalent principles, will fuffice, that is capable of fuch adjuftments as to be made correctly to defcribe an almicanther and azimuth circle, and capable of being retained in any given pofition; the ufe will appear by the following example.

Make choice of any fixed ftar, which, according to the diurnal motion, precedes the heavenly body to be obferved by a few minutes, more or lefs, as it may happen; let the inftrument be fet to an azimuth, fomewhat preceding the fixed ftar; and carefully obferve the time of the ftar's tranfit crofs the vertical wire of the telefcope; then wait till the heavenly body comes to the fame azimuth; and, when arrived within the field of view, keep gently turning the fcrew that alters the elevation of the telefcope, fo as to follow the heavenly body in altitude, keeping it interfected by the horizontal wire of the telefcope, till the body paffes the middle vertical wire, and carefully note the time of its paffage; there leave the telefcope fixed as to altitude, and releafing the horizontal motion, turn it round on its vertical axis, till you meet with fome ftar, that in a little time after you will, by rifing or falling, come to the fame almicanther; and, on its arrival, carefully note the time of its paffage a crofs the horizontal hair of the tele-fcope.

Now, from the right afcenfions and declinations of the two ftars being previoufly known, or afterwards determined from meridian obfervations, the azimuth of the firft ftar, and the altitude of the laft, at the time of their refpective paffages, may be determined by computation, which will give the altitude and azimuth of the heavenly body, for the time of the middle obfervation, when it paffed the interfection of the two wires.

The fame end may alfo be obtained by taking the obfervations in an inverted order; that is, by choofing a ftar at fuch an altitude, that the heavenly body fhall, in a com-

petent

petent time afterwards, arrive at the fame altitude, &c. but, as in thefe latitudes the alteration of azimuth is, efpecially in thofe parts that are in the neighbourhood of the zodiack, quicker than that of the altitude, I apprehend it to be eafier to follow the flower motion with the fcrew, fo as to preferve the interfeftion, than the quicker, and therefore in general to be preferred; but where it happens otherwife, or the ftars lie more conveniently, the inverfe method may be purfued.

It is true, that fome degree of dexterity and practice may be requifite in the obferver, in managing the fet fcrew, fo as to keep the objeft interfefted by the wire; but if fine fmooth fcrews, fuch as are ufed for micrometers to aftronomical quadrants, are adapted to the inftrument, as well that commanding the horizontal motion as the vertical, I apprehend the management will be perfeftly eafy and familiar to an obferver, otherwife well practifed.

It is eafy to fee, that thofe ftars are to be preferred that are neareft the heavenly body to be obferved; and that *cæteris paribus*, thofe in fuch pofitions, as rife or fall flowly, are beft for determining their altitude; and thofe that alter their azimuth flowly, are beft for determining the azimuth. To avoid intricacy in defcription, I have fuppofed only two wires, interfecting each other at a right angle, in the focus of the telefcope: but for the fake of getting a medium in fuch parts of the obfervation as depend on time, it will be proper to have not only three perpendicular wires, parallel to each other as common, but alfo three horizontal wires; the proportional diftances of which being previoufly determined by obfervation, the oblique motions may, (in parts not near the pole) be confidered as right lines.

This method is the more valuable, as it is entirely free from the knowledge of refraftions; for, fince the computation gives the real altitude from the time given, independent of refraftions, and fince the heavenly body is equally affefted by refraftion, at the fame altitude, the computed altitude of the ftar will give the real altitude of the heavenly body cleared of refraftion, which never enters the queftion; and, fince fuch ftars may be chofen as will render the time intercepted fhort, there is the lefs chance of a change of refraftion, during the time, between the middle and laft obfervation; and therefore this method will be particularly ufeful in obfervations near the horizon.

Austhorpe, April 17, 1768.   J. SMEATON.

N. B. Obfervations of this kind may be made upon the planets in the day light, by making ufe of the fun for the firft obfervation inftead of a ftar; and waiting afterwards for the appearance of the ftars.

R   OBSERVATION

OBSERVATION of a Solar Eclipse, the 4th of June, 1769, at the Observatory at Austhorpe, near Leeds, in the County of York, by J. SMEATON, F. R. S.

Read November 16, 1769.

					h.	′	″
Beginning by mean time, A. M.	-	-	-	-	6	33	1
Middle	-	-	-	-	7	26	38
End	-	-	-	-	8	20	16
Total duration	-	-	-	-	1	47	15
Digits eclipsed	-	-	-	-	6	46	0

N. B. The beginning and end of the eclipse were observed by an excellent 3½ feet treble object glass telescope, constructed by DOLLAND, with the smallest magnifier, which enlarged the diameter somewhat above 80 times. As there is no defect in quantity of light from the sun, the object glass was contracted by an aperture to 2¼ inches, and the object was perfectly sharp and distinct.

The quantity was taken by a parallel wire micrometer, upon an equatorial apparatus, which rendered it very commodious for the purpose; by which the part of the sun's diameter, remaining uneclipsed, measured at right angles to a line joining the horns, was 889, such parts as the sun's diameter, taken the same day at 1½ hour in the afternoon, measured between two parallels of declination, 2041.

The latitude I have not yet got so correctly as I expected to do; but I do not at present know, whether it exceeds or falls short of 53° 48′. The supposed longitude is 6′ of time west of Greenwich; this is deduced from its position with Wakefield, whose longitude is set down in "Maskelyne's British Mariner's Guide," as determined from an observation of the transit of Venus of 1761.

The exact knowledge at what point of the sun's circumference to look for the beginning (which was communicated to me by Mr. MASKELYNE), I found of great use, insomuch that, I believe, I saw the first discernable impression; I have, however, allowed 2″ for the time elapsed between the first perception, and the being sure it was the approach of the moon that affected that part of the sun's limb, and which latter

only

only could be noted by the clock. The firſt approach did not, however, affect the ſun's circumference by any thing like a penumbra or ſhade, but began by ſome aſperities of the moon's limb, ſeeming to thurſt themſelves into that of the ſun; and which appeared before any continued part of the ſun's circumference was cut off, or, perhaps, it might be occaſioned by the firſt approach of the moon's limb, diſturbing the little protuberances upon the ſun's circumference, occaſioned by the undulation of the air, and which, when rendered exceedingly diſtinct, appeared almoſt like the teeth of a fine ſaw. This whole appearance, to a teleſcope leſs diſtinct, would probably look like a penumbra or ſhadow.

Some time before the great ſpot was immerged, there appeared two parts of the moon's circumference more protuberent than the reſt, near the right hand horn, which ſo remarkably interrupted the regulation of the curve, that it was taken notice of by all about me, and which, doubtleſs, was occaſioned by two mountains upon the moon's ſurface, remarkably higher than the reſt, and I doubt but the ſame thing will have occurred to other obſervers.

N. B. Mr. SMEATON was prevented, by clouds, from obſerving the entrance of Venus upon the ſun the evening before.

ACCOUNT

ACCOUNT of an Observation of the right Ascension and Declination of
Mercury out of the Meridian, near his greatest Elongation, September,
1786, made by Mr. JOHN SMEATON, F. R. S. with an equatorial Mi-
crometer, of his own Invention and Workmanship; accompanied with an
Investigation of a Method of allowing for Refraction in such Kind of
Observations.

Read at the Royal Society, June 7th, 1787.

M. DE LA LANDE having announced to fome of my aftronomical friends the utility
of accurate obfervations of Mercury, at his two elongations the laft year, in
Auguft and September; I tried to get obfervations of that planet in croffing the meridian,
for fome days before and after the greateft elongation in Auguft; and though the ftate of
the atmofphere about that time was not very favourable to the purpofe, yet there was one
day that I thought unexceptionable, but could not perceive the leaft appearance of
Mercury; at which I was the rather furprifed, as I had formerly feen that planet in the
like fituation with the fame inftrument, with perfect perfpicuity *: and as I did not hear
of any one elfe having fucceeded in this obfervation, I thought it might be very poffible for the fame difappointment again to happen, with refpect to the approaching elongation in September. I judged, therefore, that it might be of fome utility to aftronomy,
if, by *any means*, a good obfervation of Mercury could be got; and alfo, that it would
be a proper fubject whereon to make trial of an inftrument for fuch purpofes, the idea
of which I had conceived, and begun to conftruct, above forty years before; but which,

---

* The inftrument mentioned is a transit, made by myfelf in the year 1763; at which time achromatic
objeCt-glaffes not having been, fo far as I knew, applied to aftronomical inftruments by others, but
having found the good effeCts thereof for other purpofes, I refolved to ufe a double achromatic objeCt-
glafs, made by Mr. DOLLAND; which being of equal aperture with the fimple objeCt-glafs, then in the
transit of the Royal Obfervatory, this telefcope I therefore efteemed to be of nearly equal validity, as to
quantity of light, with that at Greenwich, but reduced to the more commodious length of three feet fix
inche

from various avocations, I did not perfect to my satisfaction till the year 1770 *: since which time it has lain by, in hopes that something might happen, by which a full and effectual trial might be made thereof.

This instrument was originally intended as an improvement of the common wire micrometer, for the purposes of taking differences of right ascension and declination, in a more commodious and effectual manner than could be done in the method then practised of using that instrument † ; and at the same time more effectually to answer the purpose of GRAHAM's *astronomical sector*, which was contrived by him (as Dr. SMITH informs us) to supply the deficiencies of the micrometer then in use ‡.

The most necessary and fundamental improvements seemed to be ; first, that of rendering the micrometer telescope manageable upon an equatorial motion ; and, secondly, the contrivance of a stand of such solidity and stedfastness that the telescope might preserve the position in which it was placed, for a length of time, for it occurred to me, that if the telescope could be maintained at rest, or in a degree of stability superior to that of the astronomical sector, then the necessity of taking in a greater compass in declination than could be commodiously given to the field of a telescope would be the less necessary : for, instead of confining the object to a comparison with a star not differing more than a few minutes of time, or at most a quarter of an hour in right ascension,

---

* Some observations made therewith, after it was completed, I transmitted to my friend Mr. AUBERT ; the consistency of which induced him to procure a similar instrument to be made for *cometary* and such kind of observations as cannot be commodiously made in the meridian. Of this instrument the Rev. Mr. WOLLASTON has made honorable mention in Phil. Trans. Vol. LXXV. for 1785, p. 343.

† The common wire micrometer, as used by Dr. BRADLEY, and described from a paper in the Doctor's hand-writing, is given by Dr. MASKELYNE in the Philosophical Transactions, Vol. LXII. for the year 1772 ; and in addition to which I must beg leave to observe, that the telescopes then in use for the micrometer were from ten to fifteen feet long, made with wooden tubes, supported at each end upon two wooden supports, by which the telescope could be managed in altitude and azimuth ; but not to follow a celestial object in its proper motion on one centre ; which apparatus, I believe, is still remaining in the Royal Observatory.

‡ This instrument is described in SMITH's Optics, Vol. II. p. 350 ; and the original one, made by Mr. GRAHAM, was, at his death, placed in the Royal Observatory, and is mounted upon a three-legged stand of *wood.*

those

thofe comparifons could be extended to an hour or two, or even on occafion to three or four hours; there being fcarcely any part of the heavens fo devoid of ftars, of a fuitable magnitude for thefe obfervations, but that a proper one may be found within that compafs in right afcenfion, provided there is allowed the difference of a degree, either north or fouth of the objeét, in declination *.

Confidering, however, that the approaching elongation would be in the morning; and that the beft chance of feeing Mercury with this inftrument would be fome time in the twilight, between Mercury's rifing and the rifing of the fun; yet, on fuppofition of catching the planet in his paffage over the wires, there would be no chance of feeing any ftar pafs over the field, wherewith to compare him, till the following evening, which, being at leaft fourteen hours, the certain pofition of the telefcope for fo great a length of time was *almoft* more than I could reafonably hope for. To judge how far I might form an expectation, by way of a previous trial, I compared Saturn with γ Capricorni †, and found the return of the ftar to the fame place two evenings afterwards, both in right afcenfion and declination, was fo near, that I concluded I might very well expect a good obfervation of Mercury, in cafe I could get a fight of him, though the ftars wherewith he was to be compared lay at the diftance of the following evening at the fooneft.

---

* The field of the telefcope of Mr. Aubert's inftrument is two degrees; but that of the original, wherewith this obfervation was made, is 1° 17′: to gain which, the eye-glafs being immoveable (and *achromatic* to prevent the indiftinctnefs that would otherwife have taken place near the border) the magnifying power was obliged to be confiderably reduced, in refpect of what has been ufual for micrometers, that is, fo as not to exceed 20 times: in confequence, there is, therefore, no need for fo long a telefcope, this being but $34\frac{2}{3}$ inches focal length of object-glafs; but being a double achromatic, made by the late Mr. John Dolland, it is capable of as great an aperture as could be given to the fimple object-glaffes of twelve or fifteen feet telefcopes, that were then generally given to micrometers; but the pencil of light being greateft in this, it is attended with this advantage, that the fmall ftars can be feen very diftinct and in great abundance, like the modern night-glaffes: and there is in reality no need of great magnifying powers for the prefent purpofe; for the place of the wire being viewed by an eye-glafs, of about $1\frac{4}{5}$ inches focus, its place may be diftinguifhed to lefs than a 2300th part of an inch, which, on the radius of $34\frac{2}{3}$ inches, is fcarce $2''\frac{1}{2}$ of a degree; and which, as I apprehend, is nearer the truth than can reafonably be expected from inftruments out of the meridian.

† According to this obfervation ♄ preceded γ ♑ the 2d of September, at 9 h. 15′ P. M. mean time by 30′ 9″.7 MT, and with greater declination fouth than γ by 41′ 23″.

The

The micrometer is furnifhed with five horary wires, denomin‍ated in their order *a*, A, B, C, D, (B being the middle horary wire) and the two declination wires are denominated *A* and *B*, each moveable by a feparate and independent micrometer-fcrew, from the outfide of the field to the centre, and a little beyond it; fo that each wire can be moved into the place of the other when at or near the centre *.

The morning of the 23d of September, about a quarter paſt five o'clock, the air being clear and perfecdly ferene, it being then about an hour after Mercury's rifing, and near three-quarters of an hour before the rifing of the fun, I very readily found Mercury with the telefcope, and when found could eafily fee him with an opera glafs; and Mercury being then in a ſtate of very little alteration of declination, I adjuſted one of the declination wires to his apparent run, by making him traverfe the *whole* field. The obfervations were then taken as in the firſt table; and in the evening I was lucky enough to get thofe of λ Ceti and *o* Tauri, intending to repeat the whole the next morning and evening. The next morning proved cloudy, and fo continued, that I faw the planet no more; but in the evening of the 26th, I found the ſtars come again fo near the fame declination, that I was encouraged to continue the ob-fervation to fee what change would happen. It then came on bad rainy weather till the 30th, when I again repeated the obfervation, and found the ſtars to come fo near in declination, that I was fully fatisfied of the ſtability of the inſtrument, fo far at leaſt as could regard twenty-four hours: but as I was then appointed to go a journey, and could have no other ufe for the inſtrument, I locked the door of the Obfervatory, leaving the inſtrument in its pofition, that I might fee what change would happen by the time of my return; and was quite aſtoniſhed to find, on the 13th of October, that it had remained in a manner unmoved; for it had fuffered no more apparent altera-tion than what might occur by the errors of obferving, and alterations of the clocks and tranfit.

---

* In Dr. Bradley's Paper it is said, that before the *late* alterations, both the declination wires were made moveable; and that it was an improvement to make one of them fixed, and one only moveable. But however they might be immediately preceding the Doctor's time, I believe the original micrometers by Mr. Townley were with one fixed and one moveable declination wire, as I have seen one in this form among the remaining apparatus of Mr. Abraham Sharpe. In an instrument, however, fitted up for the purposes of the *equatorial micrometer*, I believe it will be found most convenient to have both those wires moveable; as by this means they not only are enabled to slide into *each other's place*, but every part of the frame of the instrument remains fixed during the whole of the observation, the two slides carrying these two wires excepted.

It

It muſt however be remarked, that, beſides that in the conſtruction of the inſtru-ment, every thing was contrived that appeared likely to give it firmneſs; it was reſted upon the *frustum* of an hexagonal pyramid of ſtone, in the founding whereof, great care was taken as to its ſolidity; and was detached from the floor for ſupporting the obſerver.

This Obſervatory at Auſthorpe I eſteem in the latitude of 53° 47′ 54″ N. and 5′ 50″ of time W. from Greenwich.

TABLE

TABLE I.

Obfervations of Mercury at his Elongation, Sept. 1786, with an Equatorial Micrometer.

Day, Object, and wires.	Hour	Time as taken by the clock	Time reduced to min. & sec.	Reduced to the middle wire	Mean of the wires	Parts of the microm.	Micrometer reduc.
Sept. 23.	A M.	M. q. bea.	′ ″	′ ″	′ ″	Rev. Pts.	Rev. Pts.
Merc. to wire *a*	5	24 3 5	24 47.5	26 34.8			
A	—	25 3 14½	25 52.3	26 34.8			
Middle wire B	—	26 2 9½	26 34.7	26 34.7	26 34.7	B 28.85	S 0.74
C	—	27 1 6	27 18	26 34.7	{ N. B. The telescope's centre		
D	—	28 1 15	28 22.5	26 34.6	{ was pointed to horary circle		
	P M.				{ Vl. 34¼ Decl. N. 7′ 48′.		
λ Ceti to B	9	15 1 27	15 28.5	15 28.5	15 28.4		
C	—	16 0 23	16 11.5	15 28.3			
o Tauri to B	—	40 1 25	40 27.5	40 27.5	40 27.4	B 8.39	N 19.72
C	—	41 0 21	41 10.5	40 27.3			
Sept. 26.							
λ Ceti to *a*	9	2 2 13	2 36.5	4 23.5			
A	—	3 2 23	3 41.5	4 23.9	4 23.8	B 16.97	N 11.14
C	—	5 0 14½	5 7.3	4 24.1			
o Tauri to A	—	28 2 21	28 40.5	29 22.9			
B	—	29 1 16	29 23	29 23	29 23	B 8.47	N 19.64
Sept. 30.							
λ Ceti to *a*	8	47 0 3½	47 1.8	48 48.8			
A	—	48 0 13	48 6.5	48 48.9			
B	—	48 3 8	48 49	48 49	48 49	B 16.97	N 11.14
C	—	49 2 4½	49 32.3	48 49.1			
o Tauri to A	9	13 0 11	13 5.5	13 47.9			
B	—	13 3 6	13 48	13 48	13 48.1	B 8.48	N 19.63
C	—	14 2 3	14 31.5	13 48.3			
α Orionis to *a*	11	41 3 12	41 51	43 38			
A	—	42 3 20½	42 53.3	43 37.7			
B	—	43 2 17	43 38.5	43 38.5	43 37.9	A 15.07	S 15.77
C	—	44 1 12	44 21	43 37.8			
D	—	45 1 20½	45 23	43 37.7			
Oct. 13.							
λ Ceti to A	7	58 3 0	58 45 .	59 27.4			
B	—	59 1 25½	59 27.7	59 27.7	59 27.5	B 16.97	N 11.14
C	8	0 0 21½	59 10.7	59 27.5			
o Tauri to C	—	25 0 20	25 10	24 26.8	24 26.8	B 8.50	N 19.61
α Orionis to *a*	10	52 2 0½	52 30.2	54 17.2			
A	—	53 2 9	53 34.5	54 16.9			
B	—	54 1 4½	54 17.2	54 17.2	54 17.1	A 15.07	S 15.77
C	—	55 0 0½	55 0.2	54 17			
D	—	56 0 10	56 5	54 17.4			
1	2	3	4	5	6	7	8

TABLE II.

For reducing the horary Wires of the Equatorial Micrometer to that of the Middle, when taken in mean Solar Time.

	Wires.	Equatorial Object.		Declination 7° 48′.	
		☉'s run	✶'s run	☉'s run	✶'s run
The 1ſt wire precedes the middle, add	a	1 46.2	1 46	1 47.3	1 47
2d ————————————	A	1 42.1	1 42	1 42.5	1 42.4
3d, or middle wire — —	B	— —	— —	— —	— —
4th is ſubſequent to the middle, ſubtract	C	1 42.9	1 42.8	1 43.3	1 43.2
5th ————————————	D	1 46.8	1 46.6	1 47.9	1 47.6
	2	3	4		6

TABLE

TABLE III.

Containing the Obfervations of Table I. reduced fo as to fhew the correct Differences of right Afcenfion and Declination between Mercury and the Stars wherewith he was compared.

1786. Date and object.	Hour.	Passage over mid. hor. wire by journ. clock.	Correction to reduce the clock to mean time.	Correct mean time of the observation.	Intervals of mean time of different observations.	Parts of micrometer from the telescope's centre.	Pts. of micr. reduced = declin. from the telescope's centre.
Sept. 23. ☿ to the mid. wire	A M 5	′ ″ 26 34.7	′ ″ — 3 59.8	h. ′ ″ 5 22 34.9		Rev. Pts. S 0.74	′ ″ S 1 8
λ Ceti to mid. wire	P M 9	15 28.4	— 4 0.1	9 11 28.3	h. ′ ″ 15 48 53.4		
o Tauri to the same	9	40 27.4	— 4 0.1	9 36 27.3	0 24 59	N 19.72	N 30 26
Sept. 26. λ Ceti to mid. wire	9	4 23.8	— 4 43.2	8 59 40.6		N 11.14	N 17 11
o Tauri to the same	9	29 23	— 4 43.2	9 24 39.8	0 24 59.2	N 19.64	N 30 18
Sept. 30. λ Ceti to mid. wire	8	48 49	— 4 50.9	8 43 58.1		N 11.14	N 17 11
o Tauri to the same	9	13 48.1	— 4 50.9	9 8 57.2	0 24 59.1	N 19.63	N 30 17
α Orion. to the same	11	43 37.9	— 4 50.8	11 38 47.1	2 29 49.9	S 15.77	S 20 20
Oct. 13. λ Ceti to mid. wire	7	59 27.5	— 6 36.4	7 52 51.3		N 11.14	N 17 11
o Tauri to the same	8	24 26.8	— 6 36.4	8 17 50	0 24 58.7	N 19.61	N 30 15
α Orion. to the same	10	54 17.1	— 6 37	10 47 40.1	2 29 50.1	S 15.77	S 20 20
1	2	3	4	5	6	7	8

TABLE

TABLE IV.

Deviations in the Direction of the Axis of the Telescope of the Equatorial Micrometer, in right Ascension and Declination in twenty Days, from the 23d of September to the 13th of October, both inclusive.

Objects observed.	3 days from the 23d to 26th.	4 days from the 26th to 30th.	7 days from the 23d to 30th.	13 days from the 30th to 13th.	17 days from the 26th to 13th.	20 days from the 23d to 13th.
λ Ceti { R. Ascension	0. not taken the 23d	too late 1.1 exact 0	too late 1.1 not taken the 23d	too soon 0.1 exact 0	too late 1.0 exact 0.	too late 1.1 not taken the 23d.
Declination	too late South 0.2 8.	too late 1.0 South 1.	too late South 1.2 9.	too soon 0.5 North 2.	too late 0.5 North 3.	too late 0.7 South 11.
o Tauri { R. Ascension	—	—	—	too soon 0.3 South 5.	—	—
Declination	—	—	—	—	—	—
α Orionis { R. Ascension	—	—	—	—	—	—
Declination	—	—	—	—	—	—

N. B. The right ascension is expressed in the integer and decimal parts of a second of time.

The declination is expressed in seconds of a degree.

Explanation of the less obvious Parts of the Tables of the Observation of
Mercury *near his Elongation,* Sept. 1786.

The third column of Table I. contains the times of obfervation as they were taken down from the half-fecond *journeyman* clock, in minutes, quarters, and *beats,* according to the following method; which was, by taking up the beat when the fecond hand came to 15, 30, 45, or 60, and then counting 30 beats repeatedly till the arrival of the objeſt at the middle of the wire it was approaching; after its arrival, the beats (or interval between two beats) being retained in memory, and the eye caſt upon the dial-plate, it was eaſily ſeen whether it was ſo many beats more than the quarter, the half, three-quarters, or the whole minute, and was ſet down accordingly. Thoſe reduced to minutes, feconds, and tenths of feconds, by allowing .3 for the quarter fecond, .5 the half, and .7 for the three-quarters of a fecond, are contained in the fourth column. The reduction of the fourth column to the fifth was by means of the auxiliary Table II.; and Mercury being then nearly ſtationary refpeſting the fun, the fun's run was uſed for the planet inſtead of that of a ſtar. The mean of each ſet of obſervations of the fifth column are carried into the ſixth.

The feventh column contains the parts of the micrometer as they were read off; to render which intelligible, it is to be noted, that the declination wire *A* travels from the upper ſide of the field of view of the teleſcope towards the centre, and ſomewhat beyond it; and upon it are taken all the objects that paſs the field of view on the upper ſide, anſwerable (by inverſion of the objeſt) to the ſouthern half of the field: and in like manner thoſe that paſs the field of view on the lower half are taken upon the wire *B,* and for the fame reafon denote a declination north. The ſcale of the micrometers of each wire begins from a point aſſumed ſomewhat without the field, and the number increaſes from thence towards the centre of the field, and continues beyond it; the integral parts are the turns of the ſcrew, and the centeſimal the diviſions of the index plate, being divided into 100 parts. The point of the ſcale, anſwerable to the centre of the field of view, having been found by obſervations on each ſcale refpectively; when the wire *A (Australis)* ſtands at 30.84, it is in the centre of the field; and when the wire *B (Borealis)* is at 28.11, it alſo cuts the fame centre. Hence the parts of the micrometer being refpectively taken from thoſe two numbers (which may therefore be called *constant* numbers) the remainder will be the diſtance of each refpective wire from the centre in parts of the micrometer. Thus, in the obſervation of *o* Tauri upon the 23d,

the

the parts are B 8.39; this taken from 28.11, leaves N 19.72, which are placed in col. 8. as the diftance, in parts of the micrometer, that o Tauri paffed north of the centre of the field of view, or axis of the telefcope.

In like manner, in the obfervation of Mercury on the 23d, the parts are B 28.85; but this being greater than the conftant number 28.11, the excefs will be .74 parts; which being the parts reaching beyond the centre, they will be fo much *south* of it, and are fet down therefore in col. 8. S 0.74: and in this manner the declinations of the reft are made out, from their refpective numbers of parts of the micrometer, and fet down in col. 8.

The numbers of the fixth column of Table I. are transferred to the third column of Table III.; and the declinations fet down in parts of the micrometer, Table I. col. 8, are transferred to col. 7. of Table III.

Col. 4. of this table, contains the corrections of the times deduced from the journey-man clock (as *per* col. 3.) to reduce it to mean time; which corrections are made out from the general account of the goings of the tranfit clock, corrected by tranfits of the fun, taken the 22d, 23d, 27th, and 30th of September, and the 12th, 13th, and 14th of October *. The journeyman clock was regularly compared at nights and mornings with the tranfit clock; and generally immediately after the obfervation. The *meridian* and *rotative* obfervatories in which the clock refpectively were, are at the diftance of 53 yards E. and W.; the comparifons were made by a feconds ftop watch †.

The numbers of the fourth column being properly applied to thofe of the third pro-duce the fifth; and which, with the fixth column, will be fufficiently explained by their titles. The parts of the micrometer in the feventh column, being reduced into minutes and feconds, are contained in col. 8. and refpectively fhew the minutes and feconds at which each object paffed to the north or fouth of the centre of the telefcope. The value of the parts of the micrometer were obtained by previous obfervations, from

---

* The transit clock was made by HINDLEY, and has a pendulum rod of *sedar* wood.

† The journeyman clock was generally fet to the transit clock on Sunday mornings; and when from home the former was fuffered to go down. The journeyman will *generally* agree with the transit clock to 2″ in 24 hours; but during the period of thefe obfervations, went remarkably well.

whence

whence the following rule was deduced : the numbers of turns and centefimal parts being confidered as integral, and divided by 1.08, the quotient will be the number of feconds. Thus, in the obfervation of *o* Tauri upon the 23d, the parts 1972, divided by 1.08, gives $1826'' = 30' 26''$; and the parts of Mercury .74, divided by 1.08 $= 68'' = 1' 8''$. Now the telefcope being fixed to one point of the heavens during the whole period of thefe obfervations, without any motion of any of the parts, the fcrews commanding the declination wires *A* and *B* excepted, we are enabled to judge of its fteadinefs to this point by the following remarks. If it varied in declination, this would be fhewn by the paffage of the fame ftar at a different diftance from the centre of the telefcope at different revolutions ; and if it varied in right afcenfion, it would be fhewn by its not paffing the horary wires at the due time, according to the acceleration of the ftars upon the mean time of the fun. Both the right afcenfion and declination may be varied by differences of refraction of the air at the fame altitude ; and the right afcenfion is further liable to be *apparently* varied, by the errors of the tranfit inftrument, the tranfit clock, the transferring of its time to the journeyman clock, the intermediate errors of the fame, and of the obfervation itfelf ; and as there paffed an interval of almoft 16 hours betwixt the paffage of Mercury over the field of view of the telefcope and that of λ Ceti, which was the neareft ftar wherewith a comparifon could be made, it will be a fatisfaction to fee, as before intimated, what variations arofe in ftill greater intervals of time.

In right afcenfion.

	h.
Thus λ Ceti upon Sept. 23, passed the horary wires at - - -	9 11 28.3
and —————— 26, —————————— - - -	8 59 40.6
λ Ceti therefore came sooner in three days by - - -	11 47.7
but ———— ought to accelerate on mean time - - -	11 47.7
———— therefore came after three days exactly to the time - -	
Again, *o* Tauri upon Sept. 23, passed the horary wires at - -	9 36 27.3
and ——————— 26, —————————— - -	9 24 39.8
o Tauri therefore came sooner after three days by - - -	11 47.5
———— ought to accelerate on mean time - - - -	11 47.7
———— therefore came too late in three days by - - - - -	.2

In declination.

	′	″
o Tauri upon Sept. 23, passed north of telescope's centre - - -	30	26
——————————— 26, ————————————————— - - -	30	18
——— therefore passed less north, or more south, than before by - -		8

In like manner every comparifon that Table III. affords is particularly fet down in Table IV. which, containing thirteen comparifons in right afcenfion and ten in declination, the greateft deviation in right afcenfion is 1″.2, and 11″ of a degree in declination. This fuppofes every error before mentioned to refide in the inftrument, and every other inftrument and obfervation, which were concerned in the refult, to be perfect; which, from the fmallnefs of the total errors, feems to indicate a degree of fteadinefs in the inftrument unexperienced or unnoticed before.

Deduction of the Position of Mercury from the preceding Observations as set down in Table III.

In right afcenfion from column 6.

	′	″
o Tauri followed λ Ceti Sept. 23. - - - - - -	24	59
—————————————— 26. - - - - - -	24	‾59.2
—————————————— 30. - - - - - -	24	59.1
—————————————— Oct. 13. - - - - - -	24	58.7
—————————————— at a mean of the four - - - -	24	59

	h.	′	″
α Orionis followed o Tauri Sept. 30. - - - - -	2	29	49.9
—————————————— Oct. 13, - - - - -	2	29	50.1
—————————————— at a mean - - - - -	2	29	50

Now Mercury preceded λ Ceti Sept. 23, - - - - -	15	48	53.4
λ Ceti preceded o Tauri by mean of four - - - - -		24	59
o Tauri preceded α Orionis by mean of two - - - -	2	29	50
Mercury therefore preceded α Orionis by - - - - -	18	43	42.4

In

In declination from column 8.

Sept. 23. A. M. Mercury passed the middle horary wire, south of its centre   -   1′ 8″

Same evening o Tauri passed the middle horary wire, north of it - - - 30 26

Therefore Mercury passed the middle horary wire more south than o Tauri by - 31 34

But Sept. 26, λ Ceti passed north of centre	17′	11″	} Diff. 13′	7″
———— o Tauri ————	30	18		
———— 30, λ Ceti ————	17	11	} —— 13	6
———— o Tauri ————	30	17		
—— Oct. 13, λ Ceti ———————	17	11	} —— 13	4
——————— o Tauri ————	30	15		

From the smallness of the above differences we may infer, that very little uncertainty in declination had attended the passage of o Tauri upon Sept. 23.

Upon Sept. 30, o Tauri passed north -	30′	17″	} Sum 54′	37″
———————— α Orionis —— south -	24	20		
Upon Oct. 13, o Tauri passed north -	30	15	} —— 54	35
———————— α Orionis —— south -	24	20		

α Orionis then at a mean passed more south than o Tauri - - - - 54 36

Mercury therefore on the 23d passed with more north declination that α Orionis - 23 2

Investigation of the Effects of Refraction.

The preceding deductions and remarks shew the consistency of the observations with themselves; yet, from the position of the telescope, it being only elevated 11°½ above the horizon *, it is necessary to examine how far the deductions above specified were capable of being affected by refraction. And in this respect it will appear, that

---

* This will readily be deduced by inspection of the celestial globe.

T   if

if it be fuppofed, there is no difference in the quantity of refraction of fuch objects as appear within the limits of tne field of view of this inftrument (which is 1° 17'), then their relative pofitions to each other will not be affected thereby: for if in fig. 1. Plate IX. we fuppofe the circle VHRO to reprefent the boundary of the field of view, HO being an horizontal and VR a vertical line, each paffing through the centre of the field at L; and if PLP denotes a part of a parallel of declination, then BLX perpendicular thereto, will be a part of an horary circle, both paffing through the fame centre. Now let $d\,\ast$ be the apparent path of a ftar, fuppofing it unaffected by refraction till it comes to the vertical line at $\ast$, and there to be lifted up by refraction in the faid vertical to L. Let $e+$ denote anothei ftar, alfo unaffected by refraction, to pafs along the different parallel of declination $e+$ till it comes to $+$; then, if it be fuppofed that the two ftars are both fituated in the fame horary circle, if at the point $+$ refraction takes place, and by hypothefis this is lifted up equally with the other, in the perpendicular $+l$, then the line $+\ast$ being drawn through the places of the two ftars, will be cotemporary and parallel to LX; and the figure $i+\ast$L being evidently a rhomboides, the two ftars, fo altered by refraction, will arrive together at the horary circle LX at the fame time, and with the fame difference of declination, as if no refraction had taken place. It is therefore only the *difference* of refraction which takes place in objects at different heights in the *same field*, that can alter their relative fituations: however, it appears neceffary to examine what this may amount to.

Let the letters in fig. 2, denote the fame things as before; to which we will add, that a, A, B, C, D, denote the parallel horary wires of the micrometer, and AA, BB, the declination wires, denored A and B in the tables: now from the celeftial globe we fhall alfo readily obtain the horary angle VLP $= 54°\frac{1}{2} = Lbc$. Let now an object pafs along the wire AA from the horizontal line at d to the vertical line at b; in this it will pafs through a difference of refraction, according as it gets more and more elevated above the horizontal line HO; and let the elevation Lb be half a degree or 30 minutes: then, according to Dr. BRADLEY's Table of Refraction *, the difference of refraction betwixt the 78th and 79th degrees of zenith diftance is $23''.6$, half of which $11''.8$, may be efteemed the difference of refraction for a difference of half a degree of altitude at $78°\frac{1}{2}$ zenith diftance, or of $11°\frac{1}{2}$ altitude: the

---

* Inferted in Dr. MASKELYNE's Obfervations, Vol. I. p. 15.

object,

Plate 9. page 138

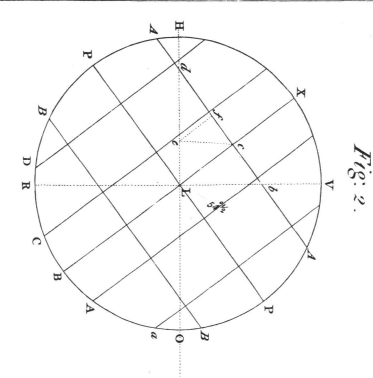

Fig: 2.

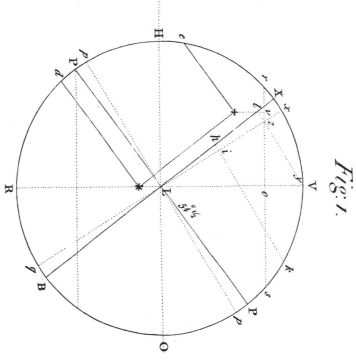

Fig: 1.

Philos. Trans. Vol. LXXVII. Tab. XII. p. 342.

object, therefore, in paffing from the horizontal line at d to the vertical line at b paffes through every difference of refraction from 0″ to 11″.8; and the queftion is, how much it is at a *medium*, that is, when it arrives at the middle wire at the point c? From this point let fall the perpendicular ce. Now, the proportion of the fides of the triangle dbL being given from conftruction, they may be taken off by a fcale viz.

$$\text{Suppofe } Lb = 174$$
$$db = 299$$
$$dL = 242$$

and affuming the fide $\quad Lb = 30$
the other fides by proportion as above $\begin{cases} db = 51.6 \\ dL = 41.7 \end{cases}$
will be

The triangles Lbc and dce are fimilar to dbL; therefore fay, as $db = 51.6 : dL = 41.7 :: Lb = 30 : Lc = 24$; and as $Lb = 30 : Lc = 24 :: dL = 41.7 : dc = 33.5$; and again, as $db = 51.6 : dc = 33.5 :: Lb = 30 : ce = 19.5$: but this will affect the declination, only in proportion of the line ef drawn parallel to LX; and it will affect the right afcenfion according to the line fc: but the triangle ecf being fimilar to the original one dbL, we fhall have $db = 51.6 : Lb = 30 :: ce = 19.5 : fc = 11.3$ for the line affecting the right afcenfion; and alfo, as $db = 51.6 : dL = 41.7 :: ce = 19.5 : ef = 15.8$ for the line affecting the declination. But the effect of difference of refraction upon the line $Lb = 30$ being only 11″.8, the refpective effects of the lines fc and ef will be in proportion; that is,

as 30′ : 11′.3 :: 11″.8 : 4″.4 for the effect in right afcenfion,
and as 30 : 15.8 :: 11.8 : 6.2 ———————— declination;

but, as it has been determined, that when the line Lb is 30 minutes, the line LC, or the correfponding declination, will be only 24 minutes; the effects of refraction above ftated will be therefore due to 24.

Correction for the Position of the Wires.

The above corrections take place on fuppofition that the feveral wires of the micrometer were ftrictly parallel to the refpective parts of the circles of declination, and horary circles in the heavens; but in the practical ufe of this inftrument it is found more convenient, on account of a ready and certain adjuftment, to place one of the

wires

wires AA or BB parallel to the apparent track of the ftar wherewith the *planetary body* is to be compared: in confequence, when the ftar ✳, fig. 1. is lifted up to L, it will not ftrictly purfue the line LP; but being lefs and lefs lifted up as it mounts higher, it will apparently fall more and more below the line LP as it afcends above the line HO, and will therefore take a courfe, fuppofe Lp. The wire PLP being therefore adjufted to agree with pLp; by conftruction of the inftrument, the wire BLX will affume the pofition qLx perpendicular to pLp. The ftar, therefore, that ran along the parallel e+ before it fuffered refraction, and at + was fuppofed to be lifted up to l, there not meeting LX will take the courfe ly, nearly parallel to Lp, and have fome diftance, as lz, to travel before it arrives at the new-placed wire Lx; and it is now proper to examine what this quantity may be.

Through the point z draw the line $r z o s$ parallel to HO, and cutting the vertical RLV in o, and let Lo be affumed$=30$; then, fince the angle XLx is fuppofed to be *minute*, the *grofs* proportions of the fides of the triangles Lyz and Lyl may be, for this purpofe, fuppofed the fame *, and the fame as Lbc, dbL, fig. 2. to which the triangle Lzo, fig. 1. will alfo be fimilar; as likewife the triangle yzo, and alfo the little triangle zlv: but making the fide lv of the triangle zlv equal to the effect of refraction in perpendicular$=11''.8$; then, to find the fide lz, the diftance run from firft to the laft fuppofed place of the wire, we need only fay, as L$b=30$: $db=51.6$ :: $lv=11''.8$: $lz=20''.3$; and this will be its value when the declination Lc, fig. 2. is 24'; but then the declination Ll or Lz, fig. 1. being greater than the perpendicular fide Lo (affumed 30') in the proportion of Lz : Lo, fay, by fimilarity of triangles converfely, as $dL=41.7$: $ab=51.6$ :: L$o=30'$: L$z=38'.2$; but as the correction before ftated of $20''.3$ is an angular error, taking place in proportion to the diftance from the centre, or the declination; for the declination given of 24' fay, as 38.2 : 24 :: $20''.3$: 13; to which, adding $4''.4$, we fhall have $17''.4$ for the whole error in right afcenfion, fuppofing it in the equator, but muft be again increafed in the proportion in which a ftar having declination is flower than a ftar in the equator; that is, it muft be increafed in the proportion of any of the numbers in

---

* I am aware, that the fuppofition of the fides of the triangles Lyz and Lyl being the fame cannot be ftrictly fo; nor can they have the fame proportions; nor are any of the lines concerned right lines that are fuppofed fuch; but affumptions *near* the truth are allowable for the correction of an error in the *greateft part*, that if uncorrected would fcarcely amount to a *grofs error*.

the

the fourth column of Table II. to the fimilar ones in col. 6. of the fame table; that is, as 1′ 46″ : 1 47″ or as 106 : 107, :: 17″.4 : 17″6 *.

As all thefe errors, arifing from difference of refraction, are in proportion of the diftance of the object from the centre of the telefcope, they will take place in proportion to the difference of declination of the two objects to be compared, whether they have paffed the field on the fame or on different fides of the centre. Now the difference of declination of Mercury and α Orionis being only 23′ 2″, and the quantities being made out for 24′ fay (rejecting the 2 feconds), as 24′ : 23′ :: 17″.4 : 16″.7, which turned into time in the run of the ftar will be 1″.1 in right afcenfion.

Say again, as 24′ : 23′ :: 6″.2 : 6″, the correction in declination. From the near equality of the lines Ll and Lz, it is evident, that no correction of *declination* is neceffary on account of the inclination of the wires, the whole difference falling in right afcenfion. As therefore Mercury paffed with 23′ 2″ more north declination than α Orionis, and paffed through a part of the *medium* that lifted him up lefs; it therefore gave him lefs north declination than it did to α, and therefore apparently diminifhed the real difference; hence 6″ muft be added to the apparent difference 23′ 2″, making it 23′ 8″, difference of declination: and as Mercury was lifted up lefs than α, he would not fo foon come to the middle wire by 1″.1 as he fhould have done, he therefore came too late by 1″.1, which muft be fubtracted from the time of Mercury's paffage the 2d of September, which will increafe the time in which he preceded α Orionis; that is, 18 h. 43′ 42″.4 increafed by 1.1 will become 18 h. 43′ 43″.5 difference of right afcenfion.

I have been the more particular in the inveftigation of this obfervation, firft of all to afcertain the degree of dependance that may be formed on an inftrument of the kind; and, fecondly, to infer fuch eafy and fimple rules, that other fimilar obfervations may be the more eafily reduced. Being therefore fatisfied of the ftability of the inftrument; if we had concluded the obfervation with that of Mercury in the morning, and of o Tauri in the evening of the 23d, then the refult from Table III. fhould have been

							h. ′ ″
Mercury passed the wires at	-	-	-	-	-	-	17 22 34.9
And o Tauri passed at	-	-	-	-	-	-	9 36 27.3
Difference of right ascension	-	-	-	-	-	-	15 13 52.4

* My friend Dr. MASKELYNE obferves, that in *ſtrictneſs* each ftar ought to have its own proper reduction, on account of *difference* of declination, which in *extreme caſes* will amount to a fenfible quantity.

which

which is the very fame as was before deduced from the mean of the whole :

					′	″
And if to Mercury's declination south of telescope's centre	·	·	·	·	1	8
We add *o* Tauris ————— north ———————	·	·	·	·	30	26
We shall have for the difference of declination	·	·	·	·	31	34

the fame as before determined.   Our obfervation would, therefore, in this cafe fimply have been, that Mercury preceded *o* Tauri in right afcenfion 15h. 13′ 52″.4 mean time, and paffed the wire with more fouth declination than *o* Tauri by 31′ 34″.

After this, *o* Tauri would have required to be compared with fome well rectified ftar by meridian inftruments; but in the prefent cafe *α* Orionis, one of Dr. MASKEYLNE's Catalogue of 34 principal ftars happened to lie fufficiently near the fame parallel of declination, to admit of *o* Tauri to be compared therewith by the fame inftrument, while pointed to the fame place of the heavens.   The operations which were fubfequent, therefore, muft be confidered as intended to fave thofe of a meridian inftrument.

Now, had our obfervation concluded with the above, then the correction would have taken place upon the difference of declination of *o* Tauri with Mercury, inftead of the ultimate one with *α* Orionis; but it muft be obferved, that whatever quantity of correction the difference of declination would occafion, it would be compenfated in the difference of refraction of *o* Tauri and *α* Orionis, when they came to be obferved on the meridian; however, in the prefent cafe it happens to be more commodious, as both can be done under one.

Preparatory then to the laying down the fimple rule for the correction of refraction, it is proper to premife, that it is evident, the lines, fig. 2 Lb, Lc, ce, ef, being in continued proportion Lb will be to ef in triplicate proportion of Lb to Lc; and that Lc will be to ef in duplicate proportion of Lb : Lc.   The difference of declination, therefore, due to 30′ difference of elevation will be as Lb to Lc fimply; but the effect of difference of refraction in declination will be lefs than the difference of declination in the proportion of Lb^2 : Lc^2; and that the effect of difference of refraction in right afcenfion will be lefs than the difference of refraction in declination in the proportion of Lb : cb fimply.

Now it has been remarked, that the elevation of the telefcope's centre above the orizon, and the horary angle VLP, will always be readily given near enough for the urpofe by the globe.   A triangle given Lbd can therefore be conftructed, and the fide

Lb

Lb being made 30′ (or any convenient aliquot part of a degree) the other fides will be found by proportion: fay then, as in the prefent cafe, $db=51.6 : dL=41.7 :: \dot{L}b=30 : Lc=24$, for the difference of declination correfponding to half a degree of altitude : fay then, as $51.6^2 : 41.7^2$, that is, as $2663 : 1739 :: 24 : 15.7 = ef$. But without troubling ourfelves with high numbers, if we take the proportion, 51.6 to 41.7 by the flide-rule *twice*, we fhall arrive at 15.7, near enough for the value of the line ef : fay then, as L$b=30 : ef$ 15.7 :: 11″.8 : 6″.2 for the refraction in declination : and as $dL=41.7 : Lb=30 :: 6″.2 : 4″.4$ for the refraction in right afcenfion, according to the *true* pofition of the wires : and for the correction of right afcenfion in the pofition of the wires, fay,

$$\text{Fig. 2.} \qquad \text{Fig. 1.}$$

As $\overbrace{Lb=30 \quad : \quad db=51.6}$ :: $\overbrace{h=11″ 8 \quad : \quad lz=20″.3}$;

and again, $dL=41.7 \cdot db=51.6 :: Lo=30′ \quad : Lz=38′.2.$

Take now Lh, fig. 1.=Lc, fig. 2.=24′, and draw the line hik, fig. 1, parallel to the line lzy, and then fay, as $Lz=38′.2 : Lh$ (or Li)=24′ :: $lz=20″.3 : hi=13″$, which $+4″.4=17″.4$ for the whole error in right afcenfion, with a declination or diftance from the centre of 24′; but as the errors both of right afcenfion and declination are in proportion to diftance from the centre, as the difference of the planet and ftar is only 23′, fay, as $24′ : 23′ :: 17″.4 : 16″.7=1″.1$ time; and for the declination, fay again, as $24′ : 23′ :: 6″.2 : 6″$ declination *.

Reduction of Mercury's Comparison with α Orionis to right Afcension and Declination.

We have laid it down, that the 23d Sept. 1786, A. M. at 5 h. 22′ 34″.9 mean time, Mercury preceded α Orionis 18 h. 43′ 43″.5, and had then a more northern declination by 23 8 .

According to Dr. MASKELYNE's Catalogue of 34 ftars, the right afcenfion of α Orionis reduced to the time when he was obferved is 85° 54′ 12″.

---

* If the comparifon had been with o Tauri, then we muft have faid,

    As $24′ : 31\frac{1}{2} :: 17″.5 : 23″=1″.5$ of time,

    and $24 : 31\frac{1}{2} :: 6.2 : 8.1$ correction for declination.

N. B. All thefe and the above proportions will be commodioufly wrought with the flide-rule.

Now, as the whole circle of the fphere makes a revolution in the time that α Orionis makes one turn, which is

h.
23 5b 4.1 then from this deduct
18 43 43.5
—————————
5 12 20.6 remains for the time that α Orionis preceded Mercury in right afcen-

fion; but if α ran the whole rotation=360° in 23 h. 56′ 4″.1, what portion of it will be run in 5 h. 12′ 20″.6=18740″.6?

But 24 h.=86,400 feconds, and 360=1,296,000 feconds:

	Time.	Time.	Of degrees.	Of degrees.

Say then, as 86400″ : 18740″.6 : : 1296000″ : 281109″=   - - 78° 5′ 9″

But, according to Dr. MASKELYNE's select Catalogue, the right ascension of α Orionis for Sept. 30, 1786, was (which add)   - 85 54 12

The right ascension of Mercury at the time of observation was therefore 163 59 21

According to Dr. MASKELYNE's select Catalogue α Orionis had decli-   ′ ″ °
   nation north, corrected for precession   -   -   - - 7 21 8.8
The sum of aberration and nutation from *Connoissance des Temps*   -   + 8.4

The correct declination north of α Orionis - ·   -   - - 7 21 17.2
To which add that Mercury passed more north   •   -   - - 23 8

Mercury's declination therefore was   -   ▾   • - 7 44 25.2

The Result.

1786, Sept. 23, A. M. } Mercury's { right ascension - - 163 59 21
at 5 h. 22′ 35″ M. T. }   { declination north - 7 44 23

DESCRIPTION

DESCRIPTION of an Improvement in the Application of the Quadrant of Altitude to a Celestial Globe, for the Resolution of Problems dependant on Azimuth and Altitude, by Mr. JOHN SMEATON, F. R. S.

Read at the Royal Society, November 20th, 1788.

PERHAPS there are few inftruments that better fulfil their defign in general, or more naturally reprefent the movements they are intended to explain and illuftrate, than the terreftrial and celeftial globe, which are alfo applied to refolve fome of the problems of the fphere, which they moft readily do. I believe, however, that whoever applies to them for the laft mentioned purpofe, will find them more defective in fome refpects than they are in others.

The difficulty that has occurred in fixing a femicircle, fo as to have a centre in the *zenith* and *nadir* points of the globe, at the fame time that the meridian is left at liberty to raife the pole to its defired elevation, I fuppofe, has induced the globe-makers to be contented with the *strip* of thin flexible brafs, called the *quadrant of altitude*; and it is well known how imperfectly it performs its office.

The improvement I have attempted, is in the application of a *quadrant of altitude*, of a more folid conftruction; which being affixed to a brafs focket of fome length, and this ground, and made to turn upon an upright fteel fpindle, fixed in the zenith, fteadily directs the *quadrant*, or rather *arc*, of *altitude* to its true *azimuth*, without being at liberty to deviate from a vertical circle to the right hand or left: by which means the azimuth and altitude are given with the fame exactnefs as the meafure of any other of the great circles.

With refpect to the horary circle, as the common application feems very convenient on account of the ready adjuftment of its index to anfwer the culmination of any of the heavenly bodies; and as I find that a circle of four inches diameter is capable of an actual and very diftinguifhable divifion into 720 parts, anfwerable to two minutes of time each, which may ferve a globe of the largeft fize; it feems that it fhould rather be *improved* than omitted; and, if inftead of a *pointer*, an index *stroke* is ufed in the fame plane with that of the divifions, the fingle minutes, and even half minutes, may be readily diftinguifhed.

U

This

This globe, though mounted merely as a model for experiment, and only nine inches in diameter, appears capable of bringing out the folution to a quarter of a degree; which, I apprehend, may be efteemed fufficient not only as a check upon numerical computation, but to come near enough to find ftars in the day-time in the field of telefcopes, which, having no equatorial motion, are only capable of direction in altitude and azimuth; but from globes of a larger fize, we may expect to come proportionably nearer.

Explanation of the Figures, Plate X.

The figures 1. and 2. being different views of the fame things, AB reprefents a line, in common to both, in the furface of the horizon, which here is of brafs.

CD, CD, are vertical lines, fuppofed to pafs through the centre of the globe in each figure; and

EFG, EFG, are portions of great circles of the globe.

Fig. 1. fuppofes the fpectator looking at the apparatus of the globe from the fouth point of the horizon; therefore the circular arch EFG, in this pofition will be a part of the *prime vertical*, and the fmall paralellogram HI is fuppofed to be a *section* of the brafs meridian, according to that vertical plane.

Fig. 2. is a view of the fame parts, the fpectator being fuppofed to look at them from the weft point of the horizon; and in this pofition HI is fuppofed to be a *portion* of the *brass meridian*. This being fixed in mind, in what follows the fame letters denote the fame parts in both figures.—KLM denotes a piece of brafs, or brafs carriage, made to fit upon the vertical part of the meridian, and capable of fliding 5° on each fide of that point, fo as adjuft to it, and to fix faft there, by means of the finger fcrew N*. This piece of brafs carries the *steel spindle* PQ which is firmly focketed into it at K, according to the dotted lines *o r*, *o r*. The axis of this fpindle is therefore capable of being fet upright upon the *zenith point*, and to maintain that pofition with a fufficient

---

* The holes reprefented in the portion of the brafs meridian (HI, fig. 2.) are fcrew holes at five degrees distance, in this quarter of the circle, into any of which the finger fcrew N is to be put, as occafion may require; the *slit* allowing fufficiently for adjustment.

Plate 10. page 146.

Fig. 1.

Fig. 2.

Fig. 3.

degree of firmnefs.—Rq, Rq, reprefents the fection of a brafs focket made to fit the fpindle, and turn round freely upon it; and when home to the fhoulder at *o o*, to turn without fhake; the focket and fpindle being a fmall matter taper, and *ground* together. On one fide of the focket is firmly fixed the arm ST, by fcrews or folder.—UW is an arch of 80 degrees, ferving inftead of the quadrant of altitude, and of the fame *fub-stance* as the meridian.   This is firmly fcrewed to the arm, and adjufted by conftruction, fo that when the *fpindle* is vertical, the face of this arch fhall make part of a vertical circle.—This arch being a portion of a circle, of the fame diameter as the brafs meridian, when its point *zero* at W refts upon the brafs horizon, its infide furface is made to agree with that of the horizon by means of a fmall thin *nib* of brafs; that being attached to the infide of the bottom of the quadrant of altitude at W, and projecting a little below it, gently bears againft the infide of the horizon, in fubftance occupying about half the *clearance* between the body of the globe and its furrounding horizon: this *nib*, feen edgeways, is fhewn at the letter X.   By this means the altitude of the object is fhewn upon the working face of the quadrant, and the quadrant s bottom fhews the azimuth upon the horizon; at the fame time the globe is free to revolve upon its axis, clear of all the circles.

The quadrant might be made complete to 90°; but as in thefe middle latitudes there is very little bufinefs for azimuths when the altitudes are above 80°, and as I judged it eligible, that the quadrant fhould be made capable of working on both fides the meridian; *that* would be prevented by the neceffary thicknefs that the circles require to give them folidity, in contradiftinction to *mathematical planes*; unlefs a part of a quadrant was cut out next the vertex to give them clearance: by this means the arch being lifted up from the fpindle, and put on the other fide of the brafs meridian for the afternoon, it will then come within 10° or 15° of the meridian; and if the ufe of this fpace fhould be wanted, it can be fupplied by reverfing the fimilar operation for the morning; and the back fide of the upper end of the quadrant at U being champered, or *bevilled* off, this will admit it to come as near to the meridian as I have mentioned.

The fteel fpindle is eafily adjufted to the *zenith*; for the globe being rectified to its *latitude*, fet the brafs carriage at liberty, bring the quadrant and meridian together, face to face, and flide the carriage, till the lower extremity of the quadrant *buts* upon the horizon, and there fcrew it faft.

It is, however, to be noted, that I have found fomething neceffary by way of *holdfaft*, to prevent the brafs meridian from fhifting its latitude, and that without confining

confining it in any other refpect.—What I have found to anfwer this purpofe is re-prefented, fig. 3. The crutch-like piece of wood ABC is fhewn as feen looking right down upon it. The circle DE is the horizontal fection of the fouth pillar of the globe. The ftrong wire pin FG, that goes through the two arms of the crutch and pillar, ferves as an axis upon which its other extremity at B is at free liberty to lift up and down, but without fhake upon the pin ; and the whole being fplit with a fine faw, from B to H, the notch BK lays hold of the underfide of the brafs meridian, and by tightening the finger fcrew LM, it firmly clips it, and retains it in any given pofition. And that it may be under no confinement *cross-ways*, the hole in the pillar is opened on both fides, as fhewn in the fection, to give it liberty of accommoda-tion ; the pin being faft in the two ends of the crutch, and turning gently in the pillar ; the whole being flender and compliant, except in point of length.

N. B. Thofe that would ufe the globe to the beft advantage to folve problems, fhould be careful to get a *just* declination, as alfo a *distinct* point to mark it; and as the circles and divifions upon the furface of the globe itfelf, are not always fuffi-ciently to be depended on for this purpofe, I have found the following expedient fully to anfwer. Choofe any plain white part of the globe's furface, anfwerable to the declination given, and with the point of a needle or protracting pin, by the help of the divifions of the brafs meridian, mark a fine point upon the blank furface of the globe, and upon this point make a dot with ink, with the fmall point of a pen, which rub off with the finger, and it will leave a fine black fpeck behind. This dot being brought to the meridian, rectify the horary index to it, and it will accurately reprefent the centre of the celeftial body whofe inveftigation is wanted.

A DESCRIPTION

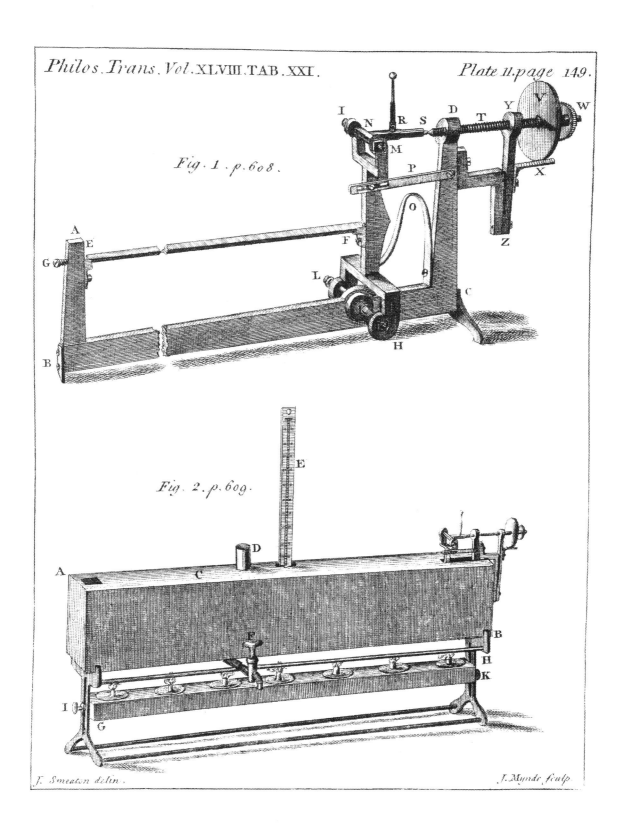

Fig. 1. p. 608.

Fig. 2. p. 609.

J. Smeaton delin.   *J. Mynde sculp.*

A DESCRIPTION of a new Pyrometer, with a Table of Experiments made therewith, by J. SMEATON, F. R. S.

See PLATE XI.

AS it may tend to illuftrate the following account, it may not be improper to make mention of the properties, that a complete inftrument, for meafuring the expanfions that metalline bodies are fubject to by heat and cold, ought to be endowed with.

And, firft, fince the quantities of thofe expanfions muft be proportionable to the length of the bar to be meafured; the longer the bar, the more fenfible the expanfion: And therefore fuch a conftruction is beft, as (*cæteris paribus*) will admit of the longeft bar.

Secondly, That the fcale, whereupon thofe minute alterations are to be meafured, ought to be, at leaft, fo large, that the fmalleft change in the length of the bar, which the inftrument is capable of being with certainty affected with, ought to be perceivable thereon.

Thirdly, As the fame change, with refpect to the pofition of the index and fcale, will enfue, upon the fuppofition, that the materials, compofing the inftrument itfelf, are expanded in a certain degree, and the bar applied to be meafured remains unchanged in its length; as if the inftrument were fuppofed to fuffer no expanfion, and the bar to be meafured were fuppofed to expand in the former degree: It is therefore neceffary, that, in the making ufe of an inftrument of this kind, the materials, of which thofe parts are compofed, upon which the meafure depends, and which may be called the bafis thereof, fhould be fubject to no expanfion or contraction during fuch trial, or that the expanfion or contraction thereof fhould be capable of being known, and accounted for.

Fourthly, That as all bodies grow ftill longer by the application of a greater degree of heat; to compare the expanfions of different bodies, we ought to have fome method of heating them in the fame degree, notwithftanding their difference of texture, fpecific gravity, &c.

Fifthly, The feveral parts, upon which the meafurements depend, ought to be fufficiently large, to be themfelves actually meafured; that not only the proportions of increafe of length in different metals, by the fame degrees of heat, may be known; but

alfo

alfo the quantities of thofe expanfions, in real meafures: Or, in other words, the proportions, that their increafe of length, between certain degrees of heat, bear to the length of the bodies: By which means, we are enabled to afcertain the changes that bodies undergo in their dimenfions by the application of any given degrees of heat.

With refpect to the firft property, this inftrument is capable of receiving a bar two feet four inches long, and might be made capable of receiving bars of a much greater length, of fome kinds of materials, but not of others; on account of the flexibility brought upon them by a degree of heat not greater than boiling water.

The meafures taken by this inftrument are determined by the contact of a piece of metal with the point of a micrometer-fcrew. The obfervation is the beft judged of by the hearing, rather than that of the fight or feeling. By this method I have found it very practicable, to repeat the fame meafurement feveral times, without differing from itfelf above one twenty-thoufandth part of an inch. This principle of determining meafures by contact is not wholly new; but has been employed on feveral occafions, as I am informed, by the late Mr. Graham: But the prefent manner of applying thereof, I believe, is fo; and the degree of fenfibility arifing therefrom exceeds any thing I have met with. As the method will eafily appear by the draught, I fhall avoid a farther defcription of it in this place *

As no fubftance has hitherto been difcovered in nature, that is perfectly free from expanfion by heat, I chofe to conftruct this inftrument in fuch a manner, that the bar, which makes the bafis of the inftrument, fhall in each experiment fuffer the fame degree of heat, as the bar to be meafured: Of confequence, the meafures taken by the micrometer are the differences of their expanfion. The expanfion then of the bafis between two given degrees of heat being once found, the abfolute expanfion of any other body, by adding or fubtracting the difference to or from the expanfion of the bafis, according as the body to be meafured expands more or lefs than the bafis, will alfo be determined.

---

* I have lately feen an instrument at Mr. Short's, made by the late Mr. Graham, for measuring the minute alterations, in length, of metal bars; which were determined by advancing the point of a micrometer-screw, till it sensibly stopped against the end of the bar to be measured. This screw being small, and very lightly hung, was capable of agreement within the 3 or 4000th part of an inch.

When

When the inftrument is made ufe of, it is immerged, together with the bar to be meafured, in a ciftern of water; which water, by means of lamps applied underneath, is made to receive any intended degree of heat, not greater than that of boiling, and thereby communicates the fame degree of heat to the inftrument, the bar, and to a mercurial thermometer immerged therein, for the purpofe of afcertaining that degree. That this may be truly the cafe, the water fhould be frequently ftirred, that there may be no difference of heat in the different parts of the water: This being done, the height of the quickfilver appearing ftationary, the contact with the fcrew of the micrometer alfo remaining the fame, for a fpace of time, it is to be fuppofed, that the heat of the three bodies will be the fame, as the heat of the water, however different they may be in fpecific gravity, &c. The whole difficulty is now reduced to this problem, viz.

To find the absolute expansion of the basis between any two given degrees of heat, not greater than that of boiling water.

For this purpofe, let there be prepared a bar of ftrait-grained white deal, or cedar; which, it is well known, are much lefs expanfible by heat than any metal hitherto difcovered: Let the bar be adapted to the inftrument in like manner as the other bars intended to be meafured; but that the foftnefs of the wood may not hinder the juftnefs of its bearings, let its ends be guarded with a bit of brafs let into the wood at the points of contact, to prevent, as much as may be, the moifture or fteam of the water from affecting the wood; let it firft be well varnifhed, and then, being wrapped round with coarfe flax from end to end; this will, in a great meafure, imbibe the vapour, before it arrives at the wood. Let the ciftern alfo be fo contrived, that the inftrument being fupported at a proper height therein, the bar to be meafured may, upon occafion, be above the cover, while the bafis remains in the water: Thus will the cover alfo be a defence againft the moifture. Let the water in the ciftern be now brought to its lower degree of heat (fuppofe at or near the freezing point), the bafis having continued long enough in the water to receive the fame degree of heat, and the wooden bar having been previoufly kept in an adjacent room, not fubject to fudden alterations of temperature by fire, or other caufes; let the bar be applied to the inftrument, and the degrees of the micrometer and the thermometer read off, and fet down: Let the wooden bar be then reftored to its former place, till the water is heated to the greater degree intended (fuppofe at or near that of boiling water); the lid being now fhut down, and the chinks ftopped with coarfe flax, to prevent the iffuing of the fteam as much as poffible, let the wooden bar be again brought forth applied to the inftrument, and the degrees of the

micrometer

micrometer and thermometer read off, as before: The difference of degrees of the micrometer, correfponding to the difference of degrees of the thermometer, will exprefs the expanfion of the bafis between thofe degrees of heat; that is, upon the fuppofition that the wooden bar was of the fame length, at the time of taking the fecond meafure, as at the firft: Indeed a meafure can hardly be taken without any lofs of time, as the whole of the inftrument, when the hot meafure is to be taken, is confiderably hotter than the wooden bar; and, in cafe of boiling water, the fteam being very repellant and active, the bar is liable to be fenfibly affected in its length, before the meafure can be taken, both by heat and moifture, which both tend to expand the bar: But as the quantity is fmall, and capable of being nearly afcertained, a wooden bar, thus applied, will anfwer the fame end as if it was unalterable by heat or moifture.

In order, therefore, to know the quantity of this alteration, let the time elapfed between the firft approach of the bar to the inftrument, and the taking of the meafure, be obferved by a fecond-watch, or otherwife; after another equal interval of time, let a fecond meafure be taken; and after a third interval, a third; and a fourth; the three differences of thefe four meafures will be found nearly to tally with three terms of a geometrical progreffion, from which the preceding term may be known, and will be the correction; which, if applied to the meafure firft taken, reduces it to what it would have been if the wooden bar had not expanded during the taking thereof.

From a few obfervations of this kind, carefully repeated, the expanfion of the bafis may be fettled; and this once done, the making experiments upon other bars will become very eafy and compendious.

The bafis of this inftrument (as well as other parts thereof) is brafs. I chofe this fubftance rather than any other whofe expanfion was greater or lefs, becaufe I found, from fome grofs experiments previoufly made, that the expanfion of brafs was nearly a medium between thofe bodies, which differ moft in their expanfion: A confiderable convenience arifes from this circumftance; becaufe as the meafures, taken in common experiments, are their difference from brafs, the dependance upon the thermometer will be lefs, as thefe differences are lefs. This precaution I have found the more neceffary, as the greateft errors that experiments made with this inftrument are fubject to, feem to be chiefly owing to the thermometer, though that which I ufed was well graduated, and good in other refpects; but this muft neceffarily happen, as the fcale and fenfibility of

the

the micrometer, when thofe metals were tried which differ moft from the bafis, were greater than that of the thermometer.

The bar of brafs which compofes the bafis is an inch broad by half an inch thick, and ftands edgeways upwards; one end is continued of the fame piece at right angles, to the height of three inches and an half, and makes a firm fupport for the end of the bar to be experimented; and the other end acts upon the middle of a lever of the fecond kind, whofe fulcrum is in the bafis; therefore the motion of the extremity of the lever is double the difference between the expanfion of the bar, and the bafis. This upper part of the lever rifes above the lid of the ciftern, fo that it and the micrometer-fcrew are at all times clear of the water. The top of the lever is furnifhed with an appendage which I call the *feeler*; it is the extremity of this piece which comes in contact with the micrometer-fcrew. The conftruction and application hereof will better appear from the draught than from many words. It hence appears, that, having the length of the lever from its fulcrum to the point of fufpenfion of the feeler, the diftance between the fulcrum and the point of contact with the bar, the inches and parts that correfpond to a certain number of threads of the micrometer, and the number of divifions in the circumference of the index-plate; the fraction of an inch expreffed by one divifion of the plate may be deduced: Thofe meafures are as follows.

			Inches.
From the fulcrum of the lever to the feeler	-	-	5.875
From the fulcrum to the plate of contact	-	-	2.895
Length of 70 threads of the fcrew	-	-	2.455
Divifions in the circumference of the index-plate 100.			

Hence the value of one divifion will be the $\frac{1}{11163}$ part of an inch: But if the fcrew be altered $\frac{1}{4}$ of one of thefe divifions, when the contact between the fcrew and feeler is well adjufted, the difference of contact (if I may fo call it) will be very perceivable to the flighteft obferver; and, confequently, $\frac{1}{2343}$ part of an inch is perceivable in this inftrument.

There is one thing ftill remains to be fpoken of, and that is, the verification of the micrometer-fcrew, which is the only part of this inftrument that requires exactnefs in the execution; and how difficult thefe are to make, perfectly good, is well known to every perfon of experience in thefe matters; that is, that the threads of the fcrew may not

X   only

only be equidiſtant, in different places, but that the threads ſhall be equally inclined to the axis in every part of the circumference.

As nearly the ſame part of the ſcrew is made uſe of in theſe experiments, the latter circumſtance is what principally needs enquiry. For this purpoſe, let a thin ſlip of ſteel, or other metal, be prepared, whoſe thickneſs is about $\frac{1}{8}$ of the diſtance of the threads: Let the edges of this thin plate be cut into ſuch a ſhape, as exactly to fit into the fixed notch in which one end of the bar is laid: Let a ſcrew paſs through the ſtandard of braſs, on which that notch is ſupported, in ſuch a manner, that the end of the bar to be meaſured that is fartheſt from the lever, may take its bearing againſt the point (or rather the ſmall hemiſpherical end) of this ſcrew: Let one of the braſs bars, uſed in the other experiments, be applied to the inſtrument, and a meaſure taken; then let the thin plate be put in between the end of the bar and the point of the ſcrew laſt mentioned, and again take the meaſure; but firſt obſerve, that the plate is put down to the notch, ſo that the ſame place of the plate may always agree with the point of the ſcrew, and, conſequently, no error may ariſe from a different thickneſs in different places of the plate: Obſerve alſo, that the whole comes to a true bearing; then advance the ſame ſcrew till the micrometer-ſcrew is puſhed backward $\frac{1}{4}$ of a revolution; again repeat the meaſure with and without the thin plate; again advance the former ſcrew, ſo as to make that of the micrometer recede another quarter of a turn, and repeat the meaſures with and without the thin plate. This method being purſued as far as neceſſary, it is evident, that, the thickneſs of the plate being always the ſame, if the difference of meaſures, taken with and without it, are not always the ſame in the different parts of a revolution of the micrometer-ſcrew, that this ſerew is not equiangular; but from the differences of the meaſures correſponding to the thickneſs of the ſame plate, in the different parts of a revolution, the errors thereof may be nearly aſſigned. For greater certainty in this examination, leſt the heat of the obſerver's body ſhould affect the bar or inſtrument during the obſervation, let the whole be immerged in the ciſtern of water, which ought to ſtand a ſufficient time before the obſervation is begun, to acquire the ſame temper as the air, which alſo ought to be in a ſettled ſtate.

In this manner I examined ſuch threads of this ſcrew as were made uſe of in the following experiments, but did not find any material errors.

The reſult of the experiments made with this inſtrument agrees very well with the proportions of expanſion of ſeveral metals given by Mr. Ellicott; which were deduced

from

from his pyrometer publifhed in the *Philofophical Tranfactions*: And, confidering the very different conftruction of the two inftruments, they abundantly tend to confirm each other.

References to the Figures.

Fig. 1. reprefents the inftrument independent of the ciftern in which it is ufed.

A B C D, is the main bar or bafis of the inftrument.

E F, is the bar to be meafured, lying in two notches; one fixed to the upright ftandard A B, the other to the principal lever H I.   The end E of the bar E F, bears againft the point of

G, a fcrew of ufe in examining the micrometer-fcrew.

The other end of the bar F bears againft a fmall fpherically protuberant bit of hard metal fixed at the fame height as G, in the principal lever H I.

K, is an arbor fixed in the bafis, which receives at each end the points of the fcrews H L, upon which the lever H L turns, and ferve as a fulcrum thereto.

O, is a flender fpring, to keep the lever in a bearing ftate againft the bar; and

P, is a check, to prevent the lever from falling forward when the bar is taken out.

N, is the feeler, fomething in the fhape of a T, fufpended, and moveable up and down upon the points of the fcrews I M, which, as well as L H, are fo adjufted; as to leave the motion free, but without fhake.

Q R, is the handle of the feeler, moveable upon a loofe joint at R; fo that, laying hold of it at Q, the feeler is moved up and down without being affected by the irregular preffure of the hand.

The extremity S of the feeler is alfo furnifhed with a bit of protuberant hard metal, to render its contact with the point of the micrometer-fcrew more perfect.

I, is

I, is the micrometer-fcrew ; V, is the divided index-plate, and W, a knob for the handle.

The micrometer-fcrew paffes through two folid fcrewed holes at D and Y.

The piece Y Z is made a little fpringy, and endeavours to pull the fcrew backwards from the hole at D ; of confequence keeps the micrometer-fcrew conftantly bearing againft its threads the fame way, and thereby renders the motion thereof perfectly fteady and gentle.

X, is the index, having divifions upon it, anfwering to the turns of the fcrew. This piece points out the divifions of the plate, as the face of the plate points out the divifors upon the index.

When the inftrument is ufed, lay hold of the knob at Q with one hand, and, moving the feeler up and down, with the other move forward the fcrew I, till its point comes in contact with the feeler ; then will the plate and index V and X fhew the turns, and parts.

Fig. 2. reprefents the inftrument immerged in its ciftern of water, ready for ufe.

A B, is the ciftern ; C, the cover ; which, when the inftrument fig. 1. is raifed upon blocks, goes on between the bar E F and the bafis B C.

D, a handle to take off the cover when hot ; E, the mercurial thermometer ; F, the cock to let out the water.

G H, a hollow piece of tin, which fupports feven fpirit lamps, which are raifed higher or lower by the fcrews I and K, in order to give the water in the ciftern a proper degree of heat.

A TABLE

A TABLE of Experiments,

By which the Numbers in Tab. I. Col. 6. Nº 1, 2, *and* 4, *were determined.*

Experiment 1. The time elapfed between approaching the bar to the inftrument and taking the firft meafure, was half a minute : Therefore the intervals between taking the fucceeding meafures was half a minute alfo. The firft meafure was 208 ; the fecond 214½ ; the third 216½ ; the fourth 217½. The differences of thefe are 6½, 2, and 1 ; which pretty well tallies with the three laft terms of the following geometrical pro greffion whofe common divifor is 2.8 ; viz. 17.7 : 6.3 : : 2.25 : .8 ; therefore, as the meafures increafed from the firft, the firft meafure being diminifhed by the firft term, viz. 208 — 17.7 = 190.3, will be the true meafure of the bar at the firft inftant of its application, before it was expanded by the heat and moifture about the inftrument.

Exp. 2. The firft meafure was 221¼ ; the fecond, 227 ; third, 230½ ; fourth, 232¾ ; whofe differences are 5¾, 3½, and 2¼ ; agreeing with the three laft terms of the following progreffion, whofe common divifor is 1.6 ; viz. 9.2 : 5.8 : : 3.6 : 2.2 ; therefore 221.25 — 9.2   212.15.

Exp. 3. The firft meafure taken was 401 ; and at that degree of heat the wooden bar did not fenfibly alter during two minutes.

Exp. 4. The firft meafure taken was 275½ ; fecond, 278½ ; third, 280¾ ; and the fourth, 282¼ : The differences are, 3, 2¼, 1½ ; agreeing with the three laft terms of the following progreffion, whofe common divifor is 1.43 ; viz. 4.4 : 3.1 : : 2.15 : 1.5 ; therefore 275.5 — 4.4 = 271.1, which is the firft meafure corrected.

Hence, as appears by Tab. I. Nº 1, 2, 3, and 4, Column 9. thefe experiments being properly reduced, agree within one divifion of the micrometer ; and the expanfion of the bafis, at a medium, is 287½ parts thereof ; correfponding to 166° of Farenheit's thermometer.

TABLE

№	Matter of the Bars.	Cold Measure.		Middle Measure.	
		Parts of Micrometer.	Deg. by Thermom.	Parts of Micrometer.	Deg. of Thermom.
1	Standard bar of white deal — — —	486.3	40½	——	——
2	— — — — — repeated	495.8	47½	——	——
3	— — — — — repeated	514.6	43½	401.0	109
4	— — — — — repeated	573.1	37	——	——
5	White glass barometer tube — —	625.0	39½	569.0	96½
6	— — — — — repeated	566.6	39	——	——
7	— — — — — repeated	557.1	47	494.7	112
8	Martial regulus of antimony — —	684.1	47	——	——
9	— — — — — repeated	751.0	39½	708.0	95½
10	Blistered steel — — —	685.7	40	——	——
11	— — — — repeated	683.8	47	638.0	113½
12	Another bar from the same gad — —	719.7	40	——	——
13	Another bar from a different gad — —	740.1	40½	——	——
14	— — — — repeated	739.9	47	694.0	114
15	Steel hardened — — —	904.5	37¾	855.0	113½
16	Dantzick iron — — —	697.5	40½	——	——
17	Another bar from the same gad —	705.8	40½	——	——
18	Another bar from a different gad —	683.9	40½	——	——
19	— — — — repeated	683.5	47	642.9	114
20	Thick wire of English iron — —	756.0	40	——	——
21	Another rod from the same piece —	694.9	40	——	——
22	Bismuth, or tin-glass — —	689.6	39	——	——
23	Copper plate hammered — —	542.1	40	——	——
24	— — — — repeated	605.0	39½	593.5	94¼
25	Another bar from the same piece —	660.0	39½	——	——
26	— — — — repeated	717.0	39¾	705.0	93½
27	Copper 8 parts, with tin 1 — —	721.0	47	715.0	111½
28	Cast brass hammered — —	581.1	39½	——	——
29	— — — — repeated	584.2	47	582.0	112½
30	The same brass unhammered — —	682.4	39½	——	——
31	Thick brass wire hard drawn — —	691.2	39½	——	——
32	— — — — repeated	694.5	47	695.5	112½
33	Brass wire softer drawn — —	741.1	39½	——	——
34	Speculum metal — —	827.7	47	828.8	112
35	Brass 16 parts, with tin 1 — —	603.7	47	602.5	111½
36	Speltre soldre, viz. brass 2, zink 1 —	615.0	47	624.2	111½
37	Fine pewter — — —	743.3	47	764.0	109½
38	Grain tin — — —	742.4	47	767.1	115
39	— — — — repeated	735.5	39	——	——
40	Soft solder, viz. lead 2, tin 1 —	681.7	47	713.0	114
41	Zink, 8 parts, tin 1, lightly hammered —	753.5	38¾	797.5	105
42	Hard lead — — —	569.5	39	——	——
43	— — — — repeated	617.7	47	664.5	110
44	Soft lead — — —	552.7	47	609.3	110
45	Zink, or speltre — —	654.9	39	——	——
46	— — — — repeated	665.6	47	723.5	114½
47	Zink hammered out half an inch a foot —	628.0	38¼	693.0	106½

TABLE CONTINUED.

Hot Measure.		Extremes.			Cold and Medium.			Mean Pts deduced from Extremes.	Irregularity.
Parts of Micrometer.	Deg. of Thermometer.	Different Parts.	Diff. Thermom.	Diff. Pts reduced to 166°.	Different Parts.	Diff. Thermometer.	Diff. Pts reduced to 166°.		
190.3	211½	—296.0	171	287.0	—	—	—	—	—
212.1	211	—283.7	163½	288.0	—	—	—	—	—
—	—	—	—	287.0	113.6	65½	114.5	—	—
271.1	211½	—302.0	174½	288.0	—	—	—	—	—
500.5	163	—124.5	123½	167.5	56.0	57	65.0	66.5	+ 1½
406.3	206	—160.3	167	159.5	—	—	—	—	—
398.0	208	—159.1	161	163.5	62.4	65	63.3	65.0	+ 1½
560.0	210	—124.1	163	126.1	—	—	—	—	—
662.5	154	— 88.5	114½	128.0	43.0	56	50.7	51.0	+ ¼
564.9	210	—120.8	170	118.2	—	—	—	—	—
566.3	210½	—117.5	163½	119.0	45.8	66½	45.5	47.4	+ 2
599.5	210	—120.2	170	117.5	—	—	—	—	—
622.3	210½	—117.8	170	115.5	—	—	—	—	—
625.0	209	—114.9	162	117.5	45.9	67	45.1	46.7	+ 1½
795.2	210	—109.3	172¼	105.3	49.0	75¾	42.7	41.9	—1
593.5	211½	—104.0	171	101.0	—	—	—	—	—
604.5	210½	—101.3	170	99.0	—	—	—	—	—
580.3	210½	—103.6	170	101.0	—	—	—	—	—
583.1	209	—100.4	162	102.5	40.6	67	40.0	40.8	+ ¾
653.6	209	—102.4	169	101.0	—	—	—	—	—
593.6	207	—101.3	167	101.0	—	—	—	—	—
606.0	211	— 83.6	172	80.7	—	—	—	—	—
511.1	205	— 31.0	165	31.2	—	—	—	—	—
582.0	142	— 23.0	102½	37.2	11.5	55¼	13.7	14.8	+ 1
627.1	203	— 32.9	163½	33.4	—	—	—	—	—
694.5	139¾	— 22.5	100	37.3	12 0	53¾	14.7	14.8	+ 1/10
703.7	209	— 17.3	162	17.7	6.0	64½	6.1	7.0	+ 1
571.8	210	— 9.3	170½	9.1	—	—	—	—	—
575.0	208½	— 9.2	161½	9.5	2.2	65½	2.2	3.8	+ 1½
673.5	210	— 8.9	170½	8.7	—	—	—	—	—
691.0	210	— 0.2	170½	0.2	—	—	—	—	—
692.3	209½	— 2.2	162½	2.2	1.0	65½	1.0	0.9	— 1/10
741.1	210	— 0.0	170½	0.0	—	—	—	—	—
826.5	208½	— 1.2	161½	1.2	1.1	65	1.1	0.5	— ½
597.6	209	— 6.1	162	6.3	1.2	64½	1.2	2.5	+ 1¼
632.5	209	+ 17.5	162	17.9	9.2	64½	9.4	7.1	—2¼
793.8	211	+ 50.5	164	51.1	20.7	62½	21.8	20.4	—1½
818.2	209	+ 75.8	162	76.8	24.7	68	24.0	30.5	+ 6½
822.6	210	+ 87.1	171	84.5	—	—	—	—	—
765.0	210	+ 83.3	163	84.7	31.3	67	30.8	33.7	+ 3
826.5	147	+ 73.0	108¼	111.8	44.0	66¼	43.9	44.4	+ ½
721.7	211	+ 152.2	172	147.2	—	—	—	—	—
753.0	211	+ 135.3	164	136.5	46.8	63	49.0	54.3	+ 5¼
680.7	211	+ 128.0	164	129.5	56.6	63	59.2	51.5	—7¾
810.0	211½	+ 155.1	172½	149.0	—	—	—	—	—
812.0	210	+ 146.4	163	149.0	57.9	67½	56.5	59.4	+ 3
744.0	150	+ 116.5	111¾	173.0	65.0	68¼	63.0	68.9	+ 6

TABLE II.

A TABLE of Expansions of Metals,

Shewing, how much a foot in length of each, grows longer by an increafe of heat correfponding to 180 degrees of Farenheit's thermometer, or to the difference between freezing and boiling water, expreffed in fuch parts whereof the unit is equal to the 10,000th part of an inch.

1	White glafs barometer-tube.	-	-	-	- 100
2	Martial regulus of antimony.	-	-	-	- 130
3	Bliftered fteel.	-	-	-	- 138
4	Hard fteel.	-	-	-	- 147
5	Iron.	-	-	-	- 151
6	Bifmuth.	-	-	-	- 167
7	Copper hammered.	-	-	-	- 204
8	Copper 8 parts, mixed with tin 1.	-	-	- 218
9	Caft brafs.	-	-	-	- 225
10	Brafs 16 parts, with tin 1.	-	-	-	- 229
11	Brafs wire.	-	-	-	- 232
12	Speculum metal.	-	-	-	- 232
13	Spelter folder, viz. brafs 2 parts, zink 1.	-	-	- 247
14	Fine pewter.	-	-	-	- 274
15	Grain tin.	-	-	-	- 298
16	Soft folder, viz. lead 2, tin 1.	-	-	- 301
17	Zink 8 parts, with tin 1, a little hammered.	-	-	- 323
18	Lead.	-	-	-	- 344
19	Zink or fpelter.	-	-	-	- 353
20	Zink hammered half an inch per foot.	-	-	- 373

P. S. It is now feveral years fince I firft obferved the very confiderable expanfion of the femi-metallic fubftance called *zink*, *spelter*, or *tootanag*; and propofed it as more fit for the purpofe of making compound pendulums, and metalline thermome-

ters,

ters, than brafs; as its expanfion feemed confiderably greater, and its confiftence, when gently hammered, not much inferior. With the fame view I have made trial of feveral other metallic compofitions, befides what is above fet down; but they all proved much inferior to zink in expanfion, and moft of them in confiftence.

It feems, that metals obferve a quite different proportion of expanfion in a fluid, to what they do in a folid ftate: for regulus of antimony feemed to fhrink in fixing, after being melted, confiderably more than zink.

DESCRIPTION

DESCRIPTION of a new Hygrometer, by Mr. JOHN SMEATON, F. R. S.

Read March 21, 1771.

HAVING fome years ago attempted to make an accurate and fenfible hygrometer, by means of a hempen cord, of a very confiderable length, I quickly found, that, though it was more than fufficiently fufceptible of every change in the humidity of the atmofphere, yet the cord was, upon the whole, in a continual ftate of lengthening. Though this change was the greateft at firft, yet it did not appear probable that any given time would bring it to a certainty; and furthermore, it feemed, that, as the cord grew more determinate in mean length, the alteration by certain differences of moifture grew lefs. Now as, confidering wood, paper, catgut, &c. there did not appear to be a likelihood of finding any fubftance fufficiently fenfible of differences of moifture, that would be unalterable under the fame degrees thereof; this led me to confider of a conftruction which would readily admit of an adjuftment; fo that, though the cord whereby the inftrument is actuated may be variable in itfelf, both as to abfolute length under given degrees of moifture, yet that, on fuppofition of a material departure from its original fcale, it might be readily reftored thereto, and in confequence, that any number of hygrometers; fimilarly conftructed, might, like thermometers, be capable of fpeaking the fame language.

The two points of heat, the more readily determinable in a thermometer, are points of freezing and boiling water. In like manner, to conftruct hygrometers which fhall be capable of agreement, it is neceffary to eftablifh two different degrees of a moifture which fhall be as fixed in themfelves, and to which we can as readily and as often have recourfe to as poffible. One point is given by making the fubftance perfectly wet, which feems fufficiently determinable; the other, that of perfect dry, but which I do not apprehend to be attainable with the fame precifion. A readinefs to imbibe wet, fo that the fubftance may be foon and fully faturated, and alfo a facility of parting with its moifture on being expofed to the fire to dry, at the fame time that neither immerfion in water, nor a moderate expofition to the warmth of the fire, fhall injure its texture, are properties requifite to the firft mover of fuch an hygrometer, that in a manner exclude all fubftances that I am acquainted with, befides hempen and flaxen threads or cords, and what are compounded thereof.

Upon

Plate 12.page 163.

Fig: 3.

Fig: 4.

Fig: 5.

Fig: 1.

Fig: 2.

1/3 of real size.

1/6 of real size.

Upon thefe ideas, in the year 1758, I conftructed two hygrometers, as near alike as poffible, in order that I might have the means of examining their agreement or difagreement on fimilar or diffimilar treatment. The interval or fcale between dry and wet, I divided into 100 equal parts, which I call the degrees of this hygrometer. The point of *o* denotes perfect dry; and the numbers increafe with the degrees of moifture to 100, which denotes perfect wet.

On comparing them for fome time, when hung up near together in a paffage or ftaircafe, where they would be very little affected with fire, and where they would be expofed to as free an air as poffible in the infide of the houfe, I found that they generally were within one degree, and very rarely differed two degrees; but, as thefe comparifons neceffarily took up fome time, and were frequently interrupted by long avocations from home, it was fome years before I could form a tolerable judgment upon them. One thing I foon obferved, not altogether to my liking, which was, that the flaxen cords, which I made ufe of, feemed to make fo much refiftance to the entry of fmall degrees of moifture, (fuch as is commonly experienced within doors in the fituation above mentioned) that all the changes were comprifed within the firft 30 degrees of the fcale; but yet, on expofing them to the warm fteam of a wafh-houfe, the index quickly mounted to 100. I was therefore defirous of impregnating the cords with fomething of a faline nature, which fhould difpofe them more forcibly to attract moifture, in order that the index might, with the ordinary changes of moifture in the atmofphere, travel over a greater part of the fcale of 100: how to do this in a regular and fixed quantity, was the fubject of many experiments, and feveral years interrupted inquiry. At laft, I tried the one hereafter defcribed, which feemed to anfwer my intention in a great meafure; and though, upon the whole, it does not appear likely that this inftrument will ever be made capable of fo accurate an agreement as mercurial thermometers are made to be, yet if we can reduce all the difagreements of an hygrometer within $\frac{1}{10}$th part of the whole fcale, it will probably be of ufe in fome philofophical inquiries, in lieu of inftruments which have not been as yet reduced to any common fcale at all.

Description of the Hygrometer.

PLATE XII.

Fig. 1 and 2, A B C, is an orthographic delineation of the whole inftrument, feen in front in its true proportion. D E is that of the profile, or the inftrument feen edgeways. F G, in both, reprefents a flaxen cord, about 35 inches long, fufpended by a turning peg F, and attached to a loop of brafs wire at A, which goes down into the

box

box cover H, which defends the index, &c. from injury, and by a glafs expofes the fcale to view.

Fig. 3 fhews the inftrument to a larger fcale, the upright part being fhortened, and the box cover removed; in which the fame letters reprefent the fame parts as in the preceding figures: G I are two loops or long links of brafs wire, which lay hold of the index K L, moveable upon a fmall ftud or centre K. The cord F G is kept moderately ftrained by a weight M, of about half a pound avoirdupoife.

It is obvious, that as the cord lengthens and fhortens, the extreme end of the index rifes and falls, and fucceffively paffes over.

N Q, the fcale, difpofed in the arch of a circle, and containing 100 equal divifions. This fcale is attached to the brafs fliding ruler Q P, which moves upon the directing piece R R, fixed by fcrews to the board, which makes the frame or bafe of the whole; and the fcale and ruler N Q P is retained in any place, nearer to or further from the centre K, as may be required by the fcrew S.

Fig. 4 reprefents, in profile, the fliding piece, and ftud I, (fig. 3) which traverfes upon that part of the index next the centre K, and which can, by the two fcrews of the ftud, be retained upon any part of the index that is made parallel, and which is done for 3 or 4 inches from the centre for that purpofe. The ftud is filed to the edges, like the fulcrum of a fcale beam, one being formed on the underfide, the other upon the upper, and as near as may be to one another. An hook, formed at the lower end of the wire loops G I, retains the index by the lowermoft edge of the ftud, while the weight M hangs by a fmall hook upon the upper edge; by thefe means the index is kept fteady, and the cords ftrained by the weight, with very little friction or burthen upon the central ftud K.

Fig. 5 is a parallelogram of brafs plate, to keep out duft, which is attached to the upper edge of the box cover H, and ferves to fhut the part of the box cover neceffarily cut away, to give leave for the wire G I to traverfe with the fliding ftud (fig. 4) nearer to, or further from, the centre of the index K, and where in (fig. 5) a is a hole, about ¼th of an inch diameter, for the wire G I to pafs through, in the rifing and falling of the index, freely without touching; b is a flit of a leffer fize, fufficient to pafs the wire, and admit the cover to come off without deranging the cord or index; $c c$ are

two

two fmall fcrews applied to two flits, by which the plate flides lengthways, in order to adapt the hole *a* to the wire G I, at any place of the ftud I upon the index K L.

Remarks on the preceding Construction.

1ft. In this conftruction the index K L, being 12 inches long, 4 inches from the extreme end, is filed fo narrow in the direction by which it is feen by the eye, that any part of thefe four inches, lying over the divifions of the fcale, becomes an index thereto. The fcale itfelf flides 4 inches, fo as to be brought under any part of the 4 inches of the index, attenuated as before mentioned.

2dly. The pofition of the directing piece R R is fo determined, as to be parallel to a right line drawn through the point *o* upon the fcale, and the centre K of the index ; confequently, as the attenuated part of the index forms a part of a radius, or right line from the fame centre, it follows, that whenever the index points to *o* upon the fcale, though the fcale is moved nearer to or further from the centre of the index, yet it produces no change in the place to which the index points.

3dly. When the divided arch of the fcale is at 10 inches from the centre (that is, at its mean diftance), then the centre of the arch and the centre of the index are coincident. At other diftances, the extremes of which are 8 or 12 inches, the centre of the divifions and centre of the index pointing thereto, not being coincident, the index cannot move over fpaces geometrically proportionable to one another in all fituations of the fcale; yet, the whole fcale not exceeding 30 degrees of a circle, it will be found on computation that the error can never be fo great as $\frac{1}{100}$th part of the fcale, or 1 degree of the hygrometer; which in this inftrument being confidered as an indivifible, the mechanical error will not be fenfible.

Choice and Preparation of the Cord.

The cord here made ufe of is of flax, and betwixt $\frac{1}{20}$th and $\frac{1}{30}$th part of an inch in diameter, which can readily be afcertained by meafuring a number of turns made round a pencil or fmall ftick. It is a fort of cord ufed in London for making nets, and is of that particular kind called by net-makers flaxen three threads laid. I do not imagine that the fabric of the cord is of the moft material confequence; but yet I

fuppofe,

suppose, when cords can be had of similar fabric, and nearly of the same size, that some small scources of variations will be avoided. In general, I look upon it that cords, the more they are twisted, the more they vary by different degrees of moisture, and the less we are certain of their absolute length; therefore those moderately twisted, I suppose, are likely to answer best.

A competent quantity of this cord was boiled in one pound avoirdupoise of water, in which was put two pennyweights troy of common salt; the whole was reduced by boiling to six ounces avoirdupoise, which was done in about half an hour. As this ascertains a given strength of brine on taking out the cord, it may be supposed that every fibre of the cord is equally impregnated with salt. The cord being dried, it will be proper to stretch it, which may be done so as to prevent it from untwisting, by tying three or four yards to two nails, against a wall, in an horizontal position, and hanging a weight of a pound or two to the middle, so as to make it form an obtuse angle. This done for a week or more in a room, will lay the fibres of the cord close together, and prevent its stretching so fast, after being applied to the instrument, as it otherwise would be apt to do.

I have mentioned the sizes and principal dimensions that I have used, as the instruments may as well be similarly constructed as otherwise; but I do not apprehend it to be very material to agree in any thing but the strength of the brine on taking the cord out of it. If the cord is adapted to the instrument some days before its first adjustment, I apprehend it will be the more settled.

Adjustment of the Hygrometer.

The box cover being taken off, to prevent its being spoiled by fire, and choosing a day naturally dry, set the instrument nearly upright, about a yard from a moderate fire, so that the cord may become dry and the instrument warm, but not so near as would spoil the finest linen by too much heat, and yet fully evaporate the moisture; there let the instrument stay till the index has got as low as it will go, now and then stroking the cord betwixt the thumb and finger downwards, in order to lay the fibres thereof close together, and thereby causing it to lengthen as much as possible. When the index has thus become stationary, which will generally happen in about an hour (more or less as the air is naturally more or less dry), by means of the peg at top, raise or depress the

index

index till it lays over point *o;* this done, remove the inftrument from the fire, and having ready fome warm water in a tea-cup, take a middling camel's-hair pencil, and dipping it in the water, gently anoint the cord till it will drink up no more, and till the index becomes ftationary, and water will no more have effect upon it, which will generally happen in about an hour. If in this ftate the index lays over the degree marked 100, all is right: if not, flack the fcrew S, and flide the fcale nearer to or further from the centre, till the point 100 comes under the index, and then the inftrument is adjufted for ufe; but, if the compafs of the flide is not fufficient to effect this, as may probably happen on the firft adjuftment, flack the proper fcrews, and move the fliding ftud I nearer to or further from the centre of the index, according as the angle formed by the index, between points of wet and dry, happeneth to be too fmall or too large for the fcale; the quantity can be eafily judged of, fo as the next time to come within the compafs of the flide of the fcale, the quantity of flide being $\frac{1}{3}$d of the length of the index, and confequently its compafs of adjuftment $\frac{1}{3}$d of the whole variable quantity. Now, as fliding the ftud I will vary the pofition of the index refpecting the point of *o,* this movement is only to be confidered as a rough or preparatory adjuftment, to bring it within the compafs of the flide of the fcale, which will not often happen to be neceffary after the firft time; but in this cafe the adjuftment muft be repeated in the fame manner, by drying and wetting as before defcribed.

It is to be remarked, that, as the cord is fuppofed to be impregnated in a given degree with common falt, and this not liable to evaporate, care muft be taken in wetting that no drops of wet be fuffered to fall from the cord: for, by the obfervance hereof, the original quantity is preferved in the cord.

Obfervations made upon Two original Hygrometers.

Thefe hygrometers were firft adjufted, after the impregnation of the cords with common falt, in February, 1770; they were kept together in a ftair-cafe till the fummer following; they were frequently obferved, and rarely found to differ more than one degree. In fummer, one of them remaining in the former place, the other was removed into a paffage through a building, which, having no doors, and the inftrument being hung fo that neither rain nor the direct rays of the fun could fall upon it, thereby it became expofed to the winds, and the free paffage of the open air. In thefe fituations the two hygrometers not only differed very greatly in quantity, but even frequently were moving different ways. They were thus continued till January, 1771, in which

space

fpace of time I obferved, that the moft ordinary place of the index was between 15°
and 25° in the open air; that at 40° the atmofphere felt very fenfibly moift, but yet it
was frequently above 60°, and more than once at 70°, or very near. I have therefore
marked the point of *o* dry; 20° the mean, 40° moift, 70° very moift, 100° wet. I do
not, however, mean thofe words (thofe of dry and wet excepted) with any other intent, than that of general direction, in like manner as thofe upon the barometer, leaving the relative degree of moifture to be judged of by the fcale.

In the month of January laft, I reftored the expofed hygrometer to its former place
in the ftair-cafe, when both inftruments were again compared together, and they rarely
differed more than 1, and never fo much as 2°. After this, they were both removed together to the out paffage; and there they agreed nearly in the fame manner,
the utmoft difference not exceeding 2°. After fome trial here, one was re-adjufted,
leaving the other hanging in its place; on reftoring the new adjufted inftrument to the
other, they now differed about 5°, the new adjufted one ftanding fo much higher. The day
following the other was re-adjufted alfo, and afterwards reftored to its place with the
former, which had been left in the out paffage; and after re-adjuftment they both
agreed to 1°. This being obferved for fome days, one of them was taken down, in
order to be packed up for London; this I have now the honour of exhibiting to the
Royal Society; and I beg to leave in the Society's houfe, that in cafe any one fhould
be defirous of having an inftrument made on the fame plan, they may have recourfe
thereto.

It appears from the foregoing obfervations, that, in the compafs of eleven months,
the cords had ftretched the value of 5°: and I alfo obferved that they both had contracted their compafs about 10°. I would therefore recommend, that an hygrometer
fhould, from its firft adjuftment, be re-adjufted at the end of three months, and again
at the end of fix months from the firft; after that, at the interval of about fix months,
to the end of two years from the beginning; and after that, I apprehend that once
a year will fuffice; the beft time of adjuftment being in the dry and warm weather of
July or Auguft: and by thefe means, I apprehend the inftrument will be always kept
within 2° of its proper point.

Refpecting the fenfibility of this inftrument, it has that in a greater degree than
its conftancy to its fcale can be depended upon, which was all that I intended; where
greater degrees of fenfibility are required, to make comparifons at fmall intervals of

time,

time, the beard of a wild oat, and other conftructions, may be ufed with advantage; this inftrument being confidered as a check upon them as to more diftant periods.

General Conclusion.

I am aware that an hygrometer, actuated by any principle of the kind here made ufe of, may not be a meafurer of the quantity of moifture, actually diffolved in, and intimately mixed with the air; but only indicates the difpofition of the air to part with, or precipitate the water contained in its fubftance; or, on the contrary, to diffolve and imbibe a greater quantity: but, as it is by feparating the effects of natural caufes that we are enabled to judge of thefe caufes, and from thence their effects when again compounded, every attempt to afcertain the operations of a fimple caufe will have its value in the fearch into nature: nor can we *a priori* determine the value of any new inftrument; for, if it fhould lead to a fingle difcovery, or even to afcertain a fingle fact, this may again lead to others of great importance, of which we might have either none, or an imperfect idea before. For my own part, I have always looked on a thick fog, and the fweating or condenfation of the water's vapours upon the walls in the infide of buildings, to be the greateft marks of a moift atmofphere: whereas I have not always found the hygrometer affected at thefe times in the higheft degree. On the contrary, at the clofe of a fine day, and the fall of the dew on the fudden approach of a froft, 1 have found the hygrometer more affected by moifture than in fome of the preceding cafes, and ftill more by a falling dew in the time of an hard froft. I juft mention thefe matters as hints for the inquiry of others; not having had length of time, fince I brought the inftrument to anfwer my intention, to make any abfolute conclufions.

I am forry I have been obliged to take up fo much compafs, to defcribe and explain a very fimple inftrument; but as I meant at the fame time to give fome idea of what is to be expected from it, I thought it more excufeable to be prolix than not fufficiently explicit.

J. SMEATON.

London, March 21, 1771.

P. S. It is to be noted, that, after each re-adjuftment, though the hygrometers would generally within a few hours come near their point, yet it was not till the next day that they could be depended on, as having come to their neareft agreement.

OBSERVATIONS on the Graduation of Astronomical Instruments; with an Explanation of the Method invented by the late Mr. HENRY HINDLEY, of York, Clock-maker, to divide Circles into any given Number of Parts. By Mr. JOHN SMEATON, F. R. S.; communicated by HENRY CAVENDISH, Esq. F. R. S. and S. A

Read November 17, 1785.

PERHAPS no part of the fcience of mechanics has been cultivated by the ingenious with more affiduity, or more defervedly fo, than the art of dividing circles for the purpofes of aftronomy and navigation. It is faid that TYCHO BRAHE and HEVELIUS laboured this part of their inftruments with their own hands; and though public rewards have at length brought forth different methods of dividing from our beft artifts, which have been communicated to the public; yet I truft it will be thought, that if any thing relative to this bufinefs remains yet behind, that may tend to furnifh the ingenious artifts, who are cultivating this field, with any new or curious idea upon the fubject, it will be well worth communicating to this learned Society: fince, if an hint, which is effentially different from any thing that (fo far as I know) the public is in poffeffion of, be once ftarted, and is purfued and worked upon by ingenious men, it is not poffible to fay, to what valuable purpofes it may be converted.

This, perhaps, will better appear by taking a fhort review of the labours of others, from the time of TYCHO BRAHE and HEVELIUS (who did not ufe telefcopic fights) to the prefent time.

The very learned, ingenious, and inventive Dr. HOOK, in his Animadverfions on the *Machina Cœlestis* of HEVELIUS, publifhed in the year 1674, has given us an elaborate defcription of a quadrant, whofe divifions were formed, and afterwards read off, by means of an endlefs fcrew, working upon the outermoft border of the limb of a quadrant; which, he fays, *does not at all depend upon the care and diligence of the instrument-maker, in dividing, graving, or numbering the divisions, for*

the

the same screw makes it from end to end; yet he has given us no account of any particular care or caution that he ufed, in preventing the fame fcrew from making larger or fmaller paces, in confequence of unequal refiftance, from a different hardnefs of the metal in different parts of the limb; nor any method of correcting or checking the fame; nor of making a fcrew, the angle of whofe threads with the axis fhall be equal in every part of the circumference; therefore the whole of this bufinefs (in which accurate *mechanists* well know confifts the whole of the difficulty) he refers to the *ingenious workman;* and, in particular, to the then celebrated Mr. TOMPION, whom, he fays, he employed to make his inftrument, and who had thereby *seen and experienced the difficulties that do occur therein:* but were any ingenious workman now to purfue the directions of Dr. HOOK, fo far as his communication thereof extends, we may conclude, that he would make a very inaccurate piece of work, far inferior in performance to what the Doctor feems to expect from it.*. But yet, I believe, it was the *first* attempt to apply the endlefs fcrew and wheel, or arch, to the purpofe of forming divifions for aftronomical inftruments; for, the Doctor fays himfelf, the perfection of this inftrument is the *way* of making the divifions; that it *excels all the common ways of division:* and in the table of contents it is intituled, *An Explication of the new Way of dividing.*

This method, however, of Dr. HOOK's was not laid afide without a very full and fufficient trial: for Mr. FLAMSTEED, in the *Prolegomena* of the third volume of *Historia Cœlestis,* informs us, that *he* contrived the fextant, wherewith his obfervations were chiefly made, from his entrance into the Royal Obfervatory in the year 1676 to the year 1689. This fextant was firft made of wood, and afterwards of iron, with a brafs limb of two inches broad, by Mr. TOMPION, at the expence of Sir JONAS MOORE; the radius thereof was 6 feet $9\frac{1}{4}$ inches; it was furnifhed with an endlefs fcrew upon its limb of 17 threads in an inch, and with telefcopic fights†. Of this inftrument Mr. FLAMSTEED gives the figure at the latter end of his *Prolegomena* before-mentioned, fufficiently large to fee the general defign; the whole being mounted upon a ftrong *polar axis* of iron, of three inches diameter.

---

* This was indeed verified in an attempt upon the fame plan by the DUC DE CHAULNES, publifhed in a Memoir of the Royal Academy of Sciences at Paris, for the year 1765.

† " —Qualem nemo, cœlo adhibens;—" Preface to Historia Cœleft. printed in one vol. 1712.

Though,

Though, in the full defcription of this inftrument, Mr. FLAMSTEED mentioned the limb's being furnifhed with *diagonal divifions*, diftinguifhing the arch to 10 feconds *; yet it is pretty clear, that it had not thefe originally upon it; but that the dependance was wholly upon the fcrew divifions, when it came out of Mr. TOMPION's hands. This one may reafonably infer from the obfervations themfelves; for the firft obfervation, fet down as taken with this inftrument, being upon the 29th of October, 1676, it was not till the 11th of September, 1677, that the column which contained the *check* angle by diagonal lines was filled up; and there was alfo a fpace of time, antecedent to that laft mentioned, wherein no obfervations are recorded as taken with this inftrument, in which time the diagonal divifions might be put on; and this will be put beyond a doubt, as he fays exprefsly, that finding, in the year 1677, that the threads of the fcrew had worn the border of the limb, he divided the limb into degrees himfelf, and drew a fet of diagonal divifions †; and then comparing the two fets of divifions together, he fometimes found them to differ a whole minute; wherefore, for correction thereof, he conftructed a new table for converfion of the revolutions and parts of the fcrew into degrees, minutes, and feconds; and which he applied in the obfervations taken in 1678.—However, notwithftanding this correction, in looking over the obfervations noted down as deduced each way, I frequently find a difference of half a minute; not unfrequently 40″; but in an obfervation of the moon, of the 9th June, 1687, I find a difference of 55″‡ which upon a radius of 6 feet 9 inches amounts to more than $\frac{1}{30}$th part of an inch.

In the year 1689, Mr. FLAMSTEED completed his *mural arc* at GREENWICH; and, in the *Prolegomena* before-mentioned, he makes an ample acknowledgement of the particular affiftance, care, and induftry of Mr. ABRAHAM SHARP; whom, in the month of Auguft, 1688, he brought into the Obfervatory, as his *amanuenfis*; and being, as Mr. FLAMSTEED tells us, not only a very fkilful mathematician, but exceedingly expert in mechanical operations §, he was principally employed in the conftruction of the mural arc; which in the compafs of fourteen months he finifhed, fo

---

* Prolegomena Hiftor. Cœleft. vol. III. p. 104.

† Ibid. p. 106. " Gradus in limbo deftribui ac diagonales duxi."

‡ Vol. I. of Hift. Cœleft. p. 343.

§ " Qui mechanices per quam expertus, pariter ac mathefeos peritus." Prolégomena, vol. III. p. 108.

greatly

greatly to the fatisfaction of Mr. FLAMSTEED, that he fpeaks of him in the higheft terms of praife *

This celebrated inftrument, of which he alfo gives the figure at the end of the *Prolegomena*, was of the radius of 6 feet 7½ inches; and in like manner as the fextant was furnifhed both with fcrew and diagonal divifions, all performed by the accurate hand of Mr. SHARP. But yet, whoever compares the different parts of the table for converfions of the revolutions and parts of the fcrew belonging to the mural arc into degrees, minutes, and feconds †, with each other, at the fame diftance from the zenith on different fides; and with their halves, quarters, &c. will find as notable a difagreement of the fcrew-work from the hand-divifions, as had appeared before in the work of Mr. TOMPION: and hence we may conclude, that the method of Dr. HOOK, being executed by two fuch mafterly hands as TOMPION and SHARP, and found defective, is in reality not to be depended upon in nice matters.

From the account of Mr. FLAMSTEED it appears alfo, that Mr. SHARP obtained the zenith point of the inftrument, or *line of collimation*, by obfervation of the zenith ftars, with the face of the inftrument on the eaft and on the weft fide of the wall ‡: and that having made the index ftronger (to prevent flexure) than that of the fextant, and thereby heavier, he contrived, by means of pullies and balancing weights, to relieve the hand that was to move it from a great part of its gravity ‡.

I have been the more particular relating to Mr. SHARP, in the bufinefs of conftructing this mural arc; not only becaufe we may fuppofe it the firft good and valid inftrument of the kind, but becaufe I look upon Mr. SHARP to have been the firft perfon that cut accurate and delicate divifions upon aftronomical inftruments; of which, independent of Mr. FLAMSTEED's teftimony, there ftill remain confiderable proofs: for, after leaving Mr. FLAMSTEED, and quitting the department above-

---

* " SHARPEIUS servus meus fidelissimus, ac omnibus quidem dotibus & facultatibus erat imbutus, " quæ ipsum tam subtili & difficili operi obeundo idoneum redderent." Prolegom. ibid.

And on finishing the instrument, he says, " Gradus describuntur sive numerantur et exsculpantur, " artificiosa manuali opera dicti domini SHARP, qui limbum partitus est, diagonales duxit, totumque " organum absolvit et perfecit: adeo ut præstantissimi quivis artifices postquam illud conspexerunt et " considerarunt, se exactius id peragere non potuisse, agnoverint." Prolegom. p. 111.

† Hist. Cœlest. vol. II. Appendix.

‡ Prolegom. p. 109.

mentioned,

mentioned *, he retired into Yorkfhire, to the village of Little Horton, near Bradford, where he ended his days about the year 1743; and where I have feen not only a large and very fine collection of mechanical tools (the principal ones being made with his own hands), but alfo a great variety of fcales and inftruments made therewith, both in wood and brafs, the divifions whereof were fo exquifite, as would not difcredit the firft artifts of the prefent times: and I believe there is now remaining a quadrant, of four or five feet radius, framed of wood, but the limb covered with a brafs plate; the fubdivifions being done by diagonals, the lines of which are as finely cut as thofe upon the quadrants at Greenwich. The delicacy of Mr. SHARP's hand will indeed permanently appear from the copper-plates in a quarto book, pub-liſhed in the year 1718, intituled, *Geometry improved by A. Sharp, Philomath.* whereof not only the geometrical lines upon the plates, but the whole of the engraving of letters and figures, were done by himſelf, as I was told by a perſon in the mathe-matical line, who very frequently attended Mr. SHARP in the latter part of his life. I therefore look upon Mr. SHARP as the firſt perſon that brought the affair of hand divifion to any degree of perfection.

Some time about the eftablifhment of the mural arc at Greenwich, the celebrated Danifh Aftronomer OLAUS ROEMER began his domeftic Obfervatory, which he finifhed in the year 1715, as we are informed by his hiftorian PETER HORREBOW, in the third volume of his works, in the tract, intituled, *Basis Astronomiæ,* publifhed in the year 1741. In this tract is the defcription of an inftrument, Table III. which not only anfwered the purpofe of the meridian arc; but, its telefcope being mounted upon a long axis, became alfo in reality what we now call a *Transit Instrument;* and which furnifhed, fo far as I have been able to learn, the firft idea thereof. One end of the axis of this inftrument being the centre of the meridian arc, and carrying its index, M. ROEMER thereby avoided the errors arifing from the plane of the mural arc not being accurately a vertical plane; and which Mr. FLAMSTEED endeavoured to check, by obferving the paffage of known ftars nearly in the fame parallel of de-clination; that is, paffing nearly over the fame part of the plane of the arc; by which he was enabled to correct or check the errors of the arc in right afcenfion. But it is the peculiar method in which ROEMER *divided* his inftruments, that occafions him here to be introduced.

---

* Mr. SHARP continued in strict correspondence with Mr. FLAMSTEED so long as he lived, as ap-peared by letters of Mr. FLAMSTEED's found after Mr. SHARP's death; many of which I have seen.

Though

Though it is a very fimple problem by which geometricians·teach how to divide a given right line into any number of parts required; yet it is ftill a much more fimple thing to fet off upon a given right line, from a point given, any number of equal parts required, where the total length is not exactly limited; for this amounts to nothing more than affuming a convenient opening of the compaffes, and beginning at the point given, to fet off the opening of the compaffes as many times in fuccef-fion, as there are equal parts required; which procefs is as applicable to the arch of a circle as it is to a right line. Of this fimple principle ROËMER endeavoured to avail himfelf.

For this purpofe M. ROEMER took two ftiff, but very fine pointed, pieces of fteel, and fixed them together, fo as to avoid, as much as poffible, every degree of fpring that would neceffarily attend long-legged compaffes, or even thofe of the fhorteft and ftiffeft kind when the points are brought near together. The diftance of the points that he chofe was about the $\frac{1}{10}$ or $\frac{1}{12}$ of an inch. This, upon a radius of $2\frac{1}{2}$ or three feet, would be about 10 minutes. With this opening, beginning at the point given, he fet off equal fpaces in fucceffion to the end of his arch, which was about 75°. Thofe were diftinguifhed·upon the limb of the inftrument by very fine points, which were referred to by a groffer divifion, the whole being properly numbered. The fub-divifion of thofe arches of 10 minutes each was performed by a double microfcope, carried upon the arm or radius of the inftrument, the common focus being furnifhed with parallel threads of fingle filk, whereof eleven being difpofed at ten equal in-tervals, comprehending together one ten-minute divifion, the diftance of the neareft threads became a very vifible fpace, anfwerable to one minute each, and therefore capable of a much further fubdivifion by eftimation.

The divifions of this inftrument were therefore, properly fpeaking, not degrees and minutes; but yet, if exactly equal, would ferve the purpofe as well, when their true value was found, which was done by comparifon with larger inftruments.

Now, if it be confidered, that in going ftep by ftep of ten minutes each, through a fpace of 75 degrees, there will be a fucceffion of 450 divifions, dependant upon each other; if it be alfo confidered, that the leaft degree of extuberance in the furface of the metal, where each new point is fet down, or the leaft hard particle (wherewith all the bafe metals feem to abound) will caufe a deviation in the firft impreffion of a taper point, and thereby produce an inequality in the divifion; it is evident, that though this inequality may be very fmall, and even imperceptible between neighbour-ing

ing divifions, yet among diftant ones, it may and will arife to fomething confiderable; which, in the menfuration of angles, will have the fame ill tendency as in near ones. Now, as M. ROEMER has given us no means of checking the diftant divifions, in refpect of each other, it is very probable that no one has followed his fteps, in cafes where great accuracy was required, in a confiderable number of divifions. For in reality this method is likely to fall far fhort of Dr. HOOK's; as Dr. HOOK's divifions being cut in a fimilar fucceffive manner, by the rotation of the fharp *edge* of the threads of a fcrew againft the exterior edge of the limb of the inftrument, a very flight degree of preffure will bring a fine fcrew of thirty threads in an inch (which he prefcribes) to touch againft an arch whofe radius is four or five feet in more than one, two, three, or four threads at once; fo that the threads fupporting one another, a fmall extuberance, or even a fmall hard particle in the metal, will be cut through or removed by the grinding or rather fawing motion of the fcrew; and which, in regard to its contact, being in reality an edge, will be much more effectual (that is, more firm) in its retention than a mere fimple point: and a repetition of the opera-tion, from the fupport of the threads to each other, will tend to mend the firft traces; whereas, in ROEMER's way, a repetition will make them worfe; for, what-ever drove forward or backward the point on firft entering, will, from the floping of the point, be confirmed and increafed in driving it deeper.

When Dr. HALLEY was chofen Aftronomer Royal (Mr. FLAMSTEED's inftruments being taken away by his executors), Mr. GRAHAM undertook to make a new mural quadrant, about the year 1725; who, uniting all that appeared valuable in the dif-ferent methods of his predeceffors, executed it with a degree of contrivance, accuracy, and precifion, before unknown: and the divifion thereof he performed with his own hand. The model of this quadrant, for ftrength, eafy management, and convenience, has been ever fince purfued as the moft perfect. What I apprehend to be peculiar in it, was the application of the arch of 96°; not only as a check upon the arc of degrees and minutes, but as fuperior thereto, being derived from the more fimple principle of *continual bisection*.

To make room for this, he has entirely rejected the fubdivifion by diagonals, and has adopted the method of the *Vernier*; but the fubdivifion of the vernier divifions he, as I apprehend for the firft time, meafured by the turns of the detached adjuft-ing fcrew, making it in fact a micrometer, by which the diftance of the *set* of the inftrument was to be meafured from the perfect coincidence of one of the actual divifions of the limb with the next ftroke of the vernier; by which means the obferva-

tion

tion could not only be read off with all the precifion that the divifion of the inftrument was capable of, but the two fets of divifions could be checked and compared
with each other.   Another thing that I apprehend to be peculiar in this inftrument,
was the more certain method of transferring and cutting the divifions, from the
original divided points, by means of the *beam-compass*, than could poffibly be done
from a *fiducial edge*, as had doubtlefs been conftantly the practice in cutting
diagonals; for, placing the fteady point of the beam-compafs in the tangent line to
that part of the arc where each divifion was to be cut, the opening of the compafs
being nearly the length of the tangent, the other point would cut the divifion in the
direction of the radius nearly ; and though in reality an arch of a circle, yet the fmall
part of it in ufe would be fo nearly a right line, as perfectly to anfwer the fame
end ; all which advantages put together, it is probable induced Mr. GRAHAM to
reject the diagonals.

Soon after the completion of this quadrant, Mr. GRAHAM undertook to execute a
zenith sector for the Rev. Dr. BRADLEY, which was fixed up at Wanftead, in
Effex, in the year 1727.   The very fimple conftruction that he adopted for this inftrument (the plumb line itfelf being the index) did not admit of the ufe of a vernier:
he therefore contented himfelf with dividing the arch of the limb of this inftrument
by primary points, as clofe as he thought neceffary, that is, by divifions of five
minutes each, and meafuring the diftance from the *set* of the inftrument to the next
point of divifion by a *micrometer* fcrew, in the conftruction of which fcrew he ufed
uncommon care and delicacy.   I have mentioned this inftrument to introduce this
obfervation ; that I think it highly probable, had Mr. GRAHAM conftructed the
great quadrant *after* the zenith fector had been fully tried, he would have rejected
not only the diagonals but the verniers alfo, as containing a fource of error within
themfelves which may be avoided by a well-made fcrew *.

It feems alfo, that Mr. GRAHAM, at the time he conftructed both thefe inftruments,
was not aware how much error could arife from the unequal expanfions of different
metals by heat or cold : for in both the radius, or frame of the inftrument, was iron,

---

* This has been found confentaneoufs to the experience of my friend Mr. AUBERT, who, on my
suggestion, has long since laid aside the use of his vernier, meafuring always by the micrometer screws
the distance between the set of the instrument, and the coincidence of the first stroke of the vernier
with the next primary division of the limb.

while the limbs were of brafs. They, however, remain in the Royal Obfervatory, perfect models, in all other refpects, of every thing that is likely to be attained in their refpective deftinations, and monuments of the fuperlative abilities of that great mechanician Mr. GRAHAM *.

Mr. GRAHAM lived till the year 1751; and during his time there were few inftruments of confequence conftructed without his advice and opinion. They were for many years done by Mr. SISSON, to whom doubtlefs Mr. GRAHAM would fully communicate his method of divifion; and from this fchool arofe that very eminent and accurate artift Mr. BIRD, whofe delicate hand, joined with great care and affiduity, enabled him ftill further to promote this branch of divifion; and which, being carried by him to a great pitch of perfection, the Commiffioners of Longitude did themfelves the credit, by an handfome reward, to induce him to publifh to the world his particular method of dividing aftronomical inftruments; which being drawn up by himfelf, in the year 1767, this matter is fully fet forth to the public: I fhall therefore only take this opportunity of obferving, that there feems to be one article in which Mr. BIRD's method may be ftill improved.

I muft here obferve, that I apprehend no quadrant, that has ever undergone a fevere examination, has been found to form a perfect arch of 90°; nor is it at all neceffary it fhould: the perfect equality of the divifions throughout the whole is the firft and primary confideration; as the proportion of error, when afcertained by proper obfervations, can be as eafily and readily applied, when the whole error of the rectangle is fifteen feconds, as when it is but five.

In this view, from the radius taken, I would compute the chord of fixteen degrees only. If I had an excellent plain fcale, I would ufe it; becaufe I fhould expect the deviation from the right angle to be lefs than if taken from a fcale of more moderate accuracy; but if not, the equality of the divifions would not be affected, though taken from any common diagonal fcale. This chord, fo prepared, I would lay off five times in fucceffion, from the primary point of o given, which would complete eighty degrees; I would then bifect each of thofe arches of 16°, as prefcribed by

---

* I have been informed, that Dr. MASKELYNE has caufed this objection to the fector to be rectified, since its removal to the Royal Obfervatory, by fubftituting an iron limb inftead of that of brafs, the points being made upon ftuds of gold.

Mr. BIRD, and laying off one of them beyond the 80th, would give the 88th degree : proceeding then by bifection, till I came to an arch of two degrees, laying that off from the 88th degree, would give the point of ninety degrees. Proceeding ftill by bifection, till I had reduced the degrees into quarters = fifteen minutes each, I would there ftop ; as from experience I know, that when divifions are over clofe, the accuracy of them, even by bifection, cannot be fo well attained as where they are moderately large. If a fpace of $\frac{4}{10}$ of an inch, which is a quarter of a degree, upon an eight feet radius, is thought too large an interval to draw the index over by the micrometer fcrew, this may be fhortened by placing another line at the diftance of one-third of a divifion on each fide of the index line, in which cafe the fcrew will never have to move the index plate more than one-third of a divifion, or five minutes ; and the perfect equality of thofe fide lines from the index line may be obtained, and adjufted to five minutes precifely, by putting each of the fide lines upon a little plate, capable of adjuftment to its true diftance from the middle one, by an adjufting fcrew.

The above hint is not confined to the chord of fixteen degrees, which prohibits the fubdivifions going lower than fifteen minutes : for if it be required to have divifions equivalent to five minutes upon the limb itfelf, then I would compute the chord of 21° 20′ *only* ; and laying it off four times from the primary point, the laft would mark out the divifion 85° 20′, pointed out by Mr. BIRD ; *fupplying the remainder* to a quadrant, from the bifected divifions as they arife, and not by the application of other *computed* chords.

In my Introduction to M. ROEMER's Method of Divifion, I have fhewn, that divifions laid off in fucceffion, by the fame opening of the compaffes, either in a right line, or in the arch of a circle, being in its idea geometrically true, and in itfelf the moft fimple of all proceffes, it has the faireft chance of being mechanically and practically exact, when cleared of the difturbing caufes. The objection therefore to his method is, the great number of repetitions, which depending upon each other in fucceffion (requiring no lefs than 540 to a quadrant, when fubdivided to ten minutes each), the fmalleft error in each, repeated 540 times, without any thing to check it by the way, may arife to a very fenfible and large amount : but in the method I have hinted, this objection will not lie : for, in the firft cafe, the affumed opening is laid off but five times ; and in the latter cafe but four times ; nor does this *repetition* arife out of the nature of the thing ; for, if you like it better, you may, in the former cafe, at once compute the chord of 64° ; and in the latter that of 85° 20′, and then

proceed

proceed wholly by bifection; supplying what is wanted to make up the quadrant, from the bifected divifions, as they arife. Mr. BIRD prefcribes this method himfelf, for the divifion of HADLEY's fectants and octants.

He, I fuppofe, was the firft, who conceived the idea of laying off chords of arches, whofe fub-divifions fhould be come at by continual bifection; but why he mixed therewith divifions that were derived from a different origin (as prefcribed in his method of dividing) I do not eafily conceive. He fays, that after he had proceeded by the bifections, from the arc of 85° 20′, the feveral points of 30°, 60°, 75°, and 90°, (all of which were laid down from the principle of the chord of 60 being equal to radius), *fell in without sensible inequality*; and fo indeed they might; but yet it does not follow that they were equally true in their places as if they had been (like the reft) laid down from the bifection from 85 20′, and therefore being the firft made, whatever error was in them, would be communicated to all connected with them, or taking their departure from them. Every heterogeneous mixture fhould be avoided where equal divifions are required. It is not the fame thing (as every good artift will fee) whether you *twice* take a meafure from a fcale as *nearly* the fame as you can, and lay them off feparately; or lay off *two* openings of the compaffes, in fucceffion, *unaltered*; for though the fame opening, carefully taken off from the fame fcale a fecond time, will doubtlefs fall into the points made by the firft, without fenfible error; yet as the floping fides of the conical cavities made by the firft point will conduct the points themfelves to the centre, there may be an error which, though infenfible to the fight, would have been avoided by the *more simple process* of laying off the opening twice, without ever altering the compaffes.

The 96 arc was, I have no doubt, invented by Mr. GRAHAM, from having perceived, in common with all preceding artifts, how very much more eafy a given line was to bifect, than to trifect, or quinquefect; and therefore the 96 arc which proceeded by bifections only (or by laying off the fame identical openings, which, as already fhewn, is ftill more fimple and unexceptionable) was wholly intended by him by way of checking the divifion of the arc of 90, which required trifections and quinquefections. But experience foon fhewed the fuperior advantage of it fo ftrongly, that the ufe of the 90 arc is now wholly fet afide, where accuracy is required; whereas the ingenuity of Mr. BIRD having fhewn a way to produce the 90 arc by bifection, when this is really purfued quite through the piece, by rejecting all divifions derived from any other origin, the 90 arc will have nothing in it to prevent its being equally unexcep-

tionable

tionable with the 96 arc; and confequently if, inftead of the 96 arc, another arc of 90 was laid down (which being upon a different radius, its divifions will ftand totally unconnected with the former), then thefe two arcs would in *reality* be a check upon each other; for being of equal validity, the mean might be taken: and if (in lieu of vernier divifions) ftrokes at the diftance of any odd number, as 7, 9, 11, or 13, are marked upon, and carried along with, the index plate; thefe will produce a check upon neighbouring divifions; and the angle may then be deduced from the medium of no lefs than four readings.

The laft works that have been made known to the public in the line of graduation (fo far as has come to my knowledge) are thofe of the very ingenious Mr. RAMSDEN, which were publifhed, by order of the Board of Longitude, in the year 1777.

From his own information, I learn, that in the year 1760, he turned his thoughts towards making an engine for dividing mathematical inftruments; and this he did in confequence of a reward offered by the Board of Longitude to Mr. BIRD, for pub-lifhing his method of graduating quadrants; for as, feveral years previous to that period, he had taken great pains to accomplifh himfelf in the art of hand-dividing, in which line Mr. BIRD had acquired his eminence, he conceived by this publication of Mr. BIRD's he fhould be reduced to the fame ftandard of performance with the reft of the trade. He, therefore, partly to fave time, and that kind of wearinefs to an ingenious mind that ever muft attend the endlefs repetition of the fame thing from morning to night; partly ftill to preferve the pre-eminence he had then gained; and partly to procure difpatch in the great increafe of demand for HADLEY's fextants and octants, in confequence of the fuccefsful application of the moon's motion to the purpofe of afcertaining the longitude at fea (which inftruments for this purpofe re-quired a degree of accuracy and certainty in the divifion, by no means neceffary thereto when applied to the fimple purpofe of obferving latitudes); I fay, for thefe confiderations, Mr. RAMSDEN determined to fet about fomething in the inftrumental way, that fhould be fufficient effectually to anfwer thefe purpofes.

Accordingly, confidering the nature of the endlefs fcrew, he fet himfelf to work upon an engine whofe divided wheel or plate was of thirty inches diameter; and though the performance of this firft effay was inferior to his expectations and wifhes, yet, with it, he was able to divide theodolites with a degree of precifion far fuperior to any thing of the kind that had been exhibited to the public.

This

This engine I myfelf faw in the fpring of the year 1768; and it appeared to me not only a very laudable attempt towards inftrumental divifions, but a very good model for the conftruction of an engine of the moft accurate kind for that purpofe. And he furthermore, at the fame time, fhewed me the model or pattern for cafting a wheel of a much larger fize, which he propofed to make upon the fame plan, and with confiderable improvements.   This being effected fome time in or about the year 1774, its accuracy was proved by making a fextant, afterwards fubjected to the examination of Mr. BIRD; who in confequence approved the method, not only as fully fufficient for the divifion of HADLEY's fextants and octants for any purpofe what-foever, but in fact for dividing any inftrument whofe radius did not exceed that of the dividing wheel, which was forty-five inches in diameter: whereupon the Board of Longitude, ever ready to encourage all endeavours that tend to the certainty and per-fection of any thing fubfervient to the purpofe of finding the longitude at fea, very properly and ufefully refolved to confer an handfome reward on Mr. RAMSDEN, for delivering a full explanation of his method of making the faid engine; which, in confequence, was publifhed by order of the Board of Longitude in the year already mentioned, viz. 1777: the defigns whereof are fo full and explicit, that whoever could not underftand that defcription, fo as to enable him to make it, would be unfit to undertake it on other accounts.

From what I have faid upon the works of the different artifts that I have mentioned, it would feem that the art of graduation was brought to that degree of perfection, that nothing material can now be added thereto: and I fhould have been apt to have thought fo myfelf, if I had not happened, in the courfe of my life, to have had a communication made to me (under the feal of fecrecy) which feems to promife yet further light and affiftance in perfecting that important art; and every impediment to the difcovery thereof being now removed, I fhall in the remainder of this effay give the cleareft defcription thereof that I am able, with fuch elucidations and improve-ments as feem to be naturally pointed out by the method itfelf.

In the autumn of the year 1741, I was firft introduced to the acquaintance of that then eminent artift, Mr. HENRY HINDLEY, of York, clock-maker;—he immedi-ately entered with me into the greateft freedom of communication, which founded a friendfhip that lafted to his death, which did not happen till the year 1771, at the age of 70.   On the firft interview, he fhewed me not only his general fet of tools, but his *engine*, at that time furnifhed with a dividing plate, with a great variety of numbers for cutting the teeth of clock wheels, &c. and alfo, for more nice and curious pur-

pofes,

pofes, furnifhed with a wheel of about thirteen inches diameter, very ſtout and ſtrong, and cut into 360 teeth; to which was applied an endleſs ſcrew, adapted thereto. The threads of this ſcrew were not formed upon a cylindric ſurface, but upon a ſolid whofe fides were terminated by arches of circles. The whole length contained fifteen threads; and as every thread (on the fide next the wheel) pointed towards the centre thereof, the whole fifteen were in contact together; and had been ſo ground with the wheel, that, to my great aſtoniſhment, I found the ſcrew would turn round with the utmoſt freedom, interlocked with the teeth of the wheel, and would draw the wheel round without any fhake or ſticking, or the leaſt ſenfation of inequality.

How long this engine might have been made before this firſt interview, I cannot now exactly afcertain: I believe not more than about a couple of years; but this I well remember, that he then fhewed me an inſtrument intended for aſtronomical pur-pofes, which muſt have been produced from the engine, and which of itſelf muſt have taken ſome time in making *.

I in reality thought myſelf much indebted to Mr. HINDLEY, for this communica-tion; but though he fhewed me his engine, and told me, that the ſcrew was cut by

---

* This instrument was of the equatorial kind; the wheel parallel to the equator, the quadrant of latitude, and semi-circle of declination, being all furnished with screws containing fifteen threads each, framed and moved in the same manner as that of the engine; the whole of which instrument was already framed, and the telescope tube in its place, which was intended to be of the inverting refracting kind, and to be furnished with a micrometer. This, however, was not completed till some years after; but, in the year 1748, I received it in London for sale. It staid with me two years, in which time I shewed it to all my mechanical and philosophical friends, amongst whom was Mr. SHORT, who afterwards pub-lished in the Philosophical Transactions, vol. XLVI. No. 493, p. 241, an account of a portable Ob-servatory, but without claiming any particular merit from the contrivance. However, the model of it differs from HINDLEY's equatorial only in the following articles. He added an azimuth circle and compass at the bottom. He omitted the endless screws, placing verniers in their stead; and at the top a reflecting telescope instead of a refractor. This instrument of HINDLEY's being afterwards returned to him unsold, I pointed out the principal deficiencies that I found therein; viz. that, in putting the in-strument into different positions, the springing of the materials was such as in some positions to amount to considerable errors. This remained with him in the same state till the year of the firſt *Transit of Venus*, viz. 1761; when it was sold to ———— CONSTABLE, Eſq. of *Burton Constable*, in *Holderness*. Mr. HINDLEY, to remedy the evil above-mentioned, applied balances to the different movements. He soon afterwards completed one, *de novo*, upon this improved plan, for his Grace the late Duke of NORFOLK. A method of balancing in much the same way, without the knowledge that it had been done before, has been fully explained, and laid before the Society, by our ingenious and worthy brother Mr. NAIRNE. Phil. Trans. vol. LXI. p. 108.

the rotation of the point of a tool, carried round upon a ftrong arm, at the diftance of the radius of the wheel from the centre of motion, which arm was carried forward by the wheel itfelf, and the wheel was put forward by an endlefs fcrew, formed upon a cylinder to a proper fize of thread, cut by his chock lathe; though he fhewed me alfo this chock lathe, and the method employed to make the threads of the fcrew *equiangular* with the axis, that is, to free the fcrew from what workmen term *drunkennefs;* and alfo fhewed me how, by the fingle fcrew of his lathe, he could cut, by means of wheel-work, fcrews of every neceffary degree of finenefs * (and, by taking out a wheel, could cut a left handed fcrew of the very fame degree of finenefs); by which means he was enabled not only to adapt his plain fcrew to the fize of the teeth of his wheel, but alfo to prevent any drunkennefs that otherwife the curved fcrew would be fubject to in confequence of being produced from the plain one; furthermore, that the fcrew and wheel, being ground together as an optic glafs to its tool, produced that degree of fmoothnefs in its motion that I obferved; and, laftly, that the wheel was cut from the dividing plate: yet, how the dividing plate was produced, he for particular reafons referved to himfelf.

Nor can he be blamed for the refervation of this one fecret; as he had, even at the time of my early acquaintance with him, a kind of forefight that from the fuperior merit of HADLEY's quadrant, a demand for that, and other inftruments for the purpofe of navigation, was likely to increafe; and that he might live to fee a public reward offered for a method of dividing them with greater accuracy and difpatch than had at that time appeared. Indeed, he had himfelf an idea, from the fatisfactory fuccefs that had attended his operations in dividing, that a fcrew and wheel, produced from his engines of one foot diameter, would have as much truth as the eight-feet quadrant at Greenwich: and though he doubtlefs greatly over-rated the accuracy of thefe miniature performances, yet it does not follow, as his methods were not confined to fo narrow a compafs, but that, his fcale of operation being proportionably enlarged, a degree of accuracy in the graduation of aftronomical inftruments may be attained in proportion.

I muft here beg leave to obferve, that there appears to me to be a natural limitation to the accuracy of inftruments, confifting of confiderable portions of a circle, fuch

---

* A machine for cutting the endlefs fcrew of Mr. RAMSDEN's engine, upon principles exactly fimilar, is fully and accurately fet forth in his Defcription of his dividing Engine above-mentioned.

such as quadrants, &c. * I do not find that the fineſt ſtroke upon the limb of a quadrant, made by Bird's own hand, if removed from its coincidence with its index, can
be replaced with any degree of certainty nearer than the 4000dth part of an inch,
though aided by a magnifying glaſs †.

A 4000dth part of an inch being then determined to be the *minium viſibile* by the
ſtrokes of an inſtrument, this will be leſs than one ſecond of a degree upon a radius of four
feet; and therefore, if the whole ſet of diviſions upon the limb could be preſerved true
to this aliquot part of an inch, the eight-feet quadrants of Greenwich might be expected
to be true to half a ſecond. How far they are from this, I do not exactly know; but I
have reaſon to think they vary from it ſome ſeconds: nay, I believe it is generally
allowed, that our largeſt quadrants, even when executed by the accurate hand of
Mr. Bird, do not exceed thoſe of a leſs ſize, by the ſame hand, in proportion to
their increaſe of radius: nor can it well be expected that they ſhould; ſince, as the
weight neceſſarily increaſes in a triplicate ratio of the radius, the great weight of the
Greenwich quadrants in moving and fixing them (as they could not be divided in their
place) may eaſily derange the framing; or even the *internal elaſticity* of the materials may give way, by a change of poſition, to ſo minute a quantity as a 4000dth
part of an inch. It therefore appears to me, that ſince the diviſions of a quadrant of
four feet radius are more than ſufficient, and even thoſe of three feet admit of all the
diſtinctneſs that in other reſpects is wanted, a three feet quadrant, in point of ſize,
is capable of all attainable exactneſs; and would be as much to be depended on as any
of thoſe now in being of eight feet. By adopting quadrants of this ſmaller ſize, we
ſhall of courſe get rid of $\frac{18}{19}$ths of the preſent weight; and conſequently of much
cumber, unhandineſs, and derangement, that muſt ariſe from that weight, as well as
the fear of totally diſcompoſing them, if ever moved out of their place.

It now comes to be time to open a principle upon which there is a proſpect of
effecting ſuch an improvement. I have ſhewn that a 4000dth part of an inch is the

---

* The zenith ſector conſiſts but of *few* degrees, with little variation of its poſition in uſing it.

† It will be to little purpoſe to attempt it with a greater power. Double microſcopes can doubtleſs be
formed to magnify objects, far leſs than a 4000dth part of an inch, to diſtinct ſurfaces; but then the
advantage of ſuch degrees of magnifying power is chiefly upon the organized bodies of nature. Let a dot,
or the fineſt point that can be made by human art, be ſo viewed, and it will appear not round, but a very
ragged irregular figure.

B b   *ultimatum*

ultimatum that we are to expect from *sight*, though aided by glaffes, when obferving the divifions of an inftrument. But in the XLVIIIth volume of the Philofophical Tranfactions (page 149 of this volume) I have fhewn the mechanifm of a new *pyrometer*, and experiments made therewith; whereby it appears, that, upon the principle of *contact*, a 24,000dth part of an inch is a very definite quantity. I remembered very well that I did not then go to the extent of what I might have afferted, being willing to keep within the bounds of *credibility* : but on occafion of the prefent fubject, I have re-examined this inftrument, and find myfelf very well authorifed to fay, that a 60,000th part of an inch, with fuch an inftrument, is a more definite and certain quantity than a 4000th part of an inch is to the *fight*, conditioned as above fpecified. The certainty of contact is, therefore, fifteen times greater than that of vifion, when applied to the divifions of an inftrument: and if this principle of certainty in contact did not take place even much beyond the limit I have now affigned, we never fhould have feen thofe exquifite mirrors for reflecting telefcopes, that have already been produced.

Thefe reflections apply immediately to my prefent fubject, as HINDLEY's method of divifion proceeds *wholly by contact,* and that of the firmeft kind; there being fcarcely need of magnifying glaffes in any part of the operation.

In the year 1748, I came to fettle in London; and the firft employment I met with was that of making philofophical inftruments and apparatus. In this fituation, my friend HINDLEY, from a principle the reverfe of jealoufy, fully communicated to me, by letter, his method of divifion; and though I was enjoined fecrecy refpecting others (for the reafons already mentioned), yet the communication was exprefsly made with an intention that I might apply it to my own purpofes.

The following are extracts from two letters, which contain the whole of what related to this fubject; and fince I have many things to obferve thereon, fo that the paraphrafe would be much greater than the text, I think it beft not to interrupt the defcription with any commentary, as perhaps his own mode of expreffion will more briefly and happily convey the general idea of the work than any I can ufe inftead of it.

MY DEAR FRIEND,   York, 14th Nov. 1748.

AS to what you was mentioning about my brother's knowing how I divided my engine plate, I will defcribe it as well as I can myfelf; but you will want a good many things to go through with it.

The

The manner is this : firft choofe the largeft number you want, and then choofe a long plate of thin brafs; mine was about one inch in breadth, and eight feet in length, which I bent like an hoop for an hogfhead, and foldered the ends together; and turned it of equal thicknefs, upon a block of fmooth-grained wood, upon my great lathe in the air (that is, upon the end of the mandrel) : one fide of the hoop muft be rather wider than the other, that it may fit the better to the block, which will be a fhort piece of a cone of a large diameter: when the hoop was turned, I took it off, cut, and opened it ftraight again.

 The next ftep was to have a piece of fteel bended into the form as *per margin* *; which had two fmall holes bored in it, of equal bignefs, one to receive a fmall pin, and the other a drill of equal fize. I ground the holes after they were hardened, to make them round and fmooth. The chaps formed by this fteel plate were as near together as juft to let the long plate through. Being open at one end, the chaps fo formed would fpring a little, and would prefs the long plate clofe, by fetting in the *vifc.* Then I put the long plate to a right angle to the length of the fteel chaps, and bored one hole through the long plate, into which I put the fmall pin; then bored through the other hole; and by moving the fteel chaps a hole forward, and putting in the pin in the laft hole, I proceeded till I had divided the whole length of the plate.

The next thing was to make this into a circle again. After the plate was cut off at the end of the intended number, I then proceeded to join the ends, which I did thus : I bored two narrow fhort brafs plates † as I did the long one, and put one on the infide, and the other on the outfide of the hoop, whofe ends were brought together; and put two or three turned fcrew pins, with flat heads and nuts to them, into each end, which held them together till I rivetted two little plates, one on each fide of the narrow plate, on the outfide of the hoop. Then I took out the fcrews, and turned my block down, till the hoop would fit clofe on; and by that means my right line was made into an equal divided circle of what number I pleafed.

---

* The figure is considerably lefs than the real tool fhould be.

† Thefe I fhall hereafter diftinguifh by the name of *faddle plates.*

The

The engine plate was fixed on the face of the block, with a ſteel hole fixed before it, to bore through; and I had a point that would fall into the holes of the divided hoop; ſo by cutting ſhorter, and turning the block leſs, I got all the numbers on my plate.

I need not tell you, that you get as many prime numbers as you pleaſe; nor that the diſtance of the holes in the ſteel chaps muſt be proportioned to the length of the hoop.

You may aſk my brother what he knows about my method of dividing; but need not tell him what I have ſaid about it; for I think neither *he* nor *John Smith* know ſo much as I have *told you*, though I believe they got ſome knowledge of it in general terms *.——I deſire you to keep the method of dividing to *yourſelf*, and conclude with my beſt wiſhes,

And am, dear Sir, yours, &c.

HEN. HINDLEY.

Though the above letter was in itſelf very clear and explicit as to the general traces of the method, yet ſome doubts occurring to me, a further explanation became neceſſary. A copy of my letter not being preſerved, the purport of it may be inferred from the anſwer, which was as follows:

DEAR FRIEND,   York, 13th March, 1748-9.

I THINK in your laſt you ſeem to be apprehenſive of ſome difficulties in drilling the hoop for dividing: Firſt, that the centre of the hole in the hoop might not be preciſely in the centre of the hole of the ſteel chaps, it was drilled in; but if I deſcribed fully to you the method I uſed, I can ſee no danger of error there: for my chaps were very thick, and the two correſponding holes were a little conical, and ground with a ſteel pin; firſt one pair, and then the other, alternately, till the pin would go the ſame depth into each. Then for drilling the hoop, I took any common drill that would paſs through, and bore the hole. After that I took a five-ſided

---

* The perſons here referred to were both bred with him. His brother, Mr. ROGER HINDLEY, who has many years followed the ingenious profeſſion of a watch-cap maker in London, was ſo much younger as to be an apprentice to him. Mr. JOHN SMITH, now dead, had ſome years paſt, the honour to work in the inſtrument way, under the direction of the late Dr. DEMAINBRAY, for his preſent MAJESTY.

broach,

broach, which opened the hole in the brafs betwixt the fteel chaps, but would not touch the fteel; fo confequently the centre of the holes in the brafs muft be concentric with the holes in the chaps; and for alterations by air, heat, cold, &c. I was not above two or three hours in drilling a row of holes, as far as I remember.

2dly, For drilling, in a right line, I had a thin brafs plate, faftened between the fteel chaps, for the edge of the hoop to bear againft, whilft I thruft it forward from hole to hole. What you propofe of an iron frame with a lead outfide, will be better than my wooden block; but confidering the little time that paft, betwixt transferring the divifions of the hoop to the divifions of my dividing plate, I did not fuffer much that way. It was when I drilled the holes in my dividing plate that I ufed a frame for drilling, which had one part of it that had a fteel hole, that in lying upon the plane of the dividing plate was fixed faft in its place for the point of the drill to pafs through: then, at the length of the drill, there was another piece of fteel, with a hole in it, to receive the other end of the drill to keep it at right-angles to the plane of the plate. This piece was a fpring, which bended at the end, where it was faftened to the frame of the lathe, at about eighteen inches from the end of the drill; fo it pufhed the drill through with any given force the drill would bear: and though that end of the drill moved in the arch of a circle, it was a very fmall part of it, being no more than equal to the thicknefs of the dividing plate.

My good wifhes conclude me yours,

HEN. HINDLEY.

Whoever attentively confiders the communication contained in the above letters will fee, that more happy expedients could not have been devifed to procure a fet of divifions, where there fhould be the moft exact equality among *neighbours;* and which, for the purpofes of clock-making, is the principal thing to be wifhed for. But herein, as in M. ROEMER's method, there were no means of checking the diftant divifions, which run on to 360: now fuch a check, when the expanfion of metals is confidered, and particularly the difference of expanfion between brafs and fteel, feems abfolutely neceffary for the purpofe of divifions upon inftruments, where the accurate menfuration of large angles is required, as much as the equality of neighbouring divifions *.

---

* The ingenious Mr. STANCLIFFE (some years a workman of HINDLEY's) has suggested, that the difference of expansion between the steel chaps and the brass hoop may be avoided by making the chaps of brass also, with hard steel holes set separately therein, somewhat similar to the jewelled holes of watches.

With

With this view the invention of this ingenious perſon ſuggeſted to him the thought of making his curved ſcrew to lay hold of fifteen teeth or degrees together: this, in effect, becomes a pair of compaſſes, 24 removes of which complete the whole circle, and produce 24 checks in the circumference: and whoever conſiders the very exquiſite degree of truth that reſults from the grinding of ſurfaces in contact, as already noticed, muſt expect a very great degree of rectification of whatever errors might ſubſiſt in the wheel after its firſt cutting.

What degree of truth it might in reality be capable of upon its firſt production and adjuſtment, is not now to be aſcertained, he never having uſed it for the graduation of any capital inſtrument. Thoſe that he made with a view to an accurate meaſure of angles, he always made with a ſcrew and wheel, or parts of circles cut by his engine into teeth, and ground together as before-mentioned; but I have reaſon to think, that its performance, if put to a ſtrict teſt, was never capable of that accuracy that he himſelf ſuppoſed it to have.

The method itſelf, however, from its ſimplicity, and eaſe of execution, ſeems to me to be a foundation for every thing that can be expected in truth of graduation; and in conſequence for reducing inſtruments to the leaſt ſize that is capable of bringing out all that can be expected from the largeſt; when it ſhall, like manual diviſion, have received thoſe advantages that the joint labours of the moſt ingenious men can beſtow upon it. That I may not appear to be without grounds for my expectations, I will beg leave to propoſe, what near forty years occaſional contemplation has ſuggeſted to me on the ſubject; and as I can deſcribe the proceſs I would purſue, where different from HINDLEY's, in fewer words than I could make out a regular criticiſm upon his letters, I will immediately proceed to the deſcription of it.

Proposed Improvements of HINDLEY's Method.

I would recommend the number of parts into which the circle is to be reduced to be 1440; that is 4 times 360; which diviſions will therefore be quarters of a degree; the diſtances of the holes in the chaps will therefore, to a three-feet radius, be $\frac{157}{1000}$ of an inch nearly; that is, between the one-ſixth and one-ſeventh of an inch diſtance centre and centre.

Having

Having provided myfelf with a ftout mandrel, or arbor, for a *chock lathe*, properly framed, that would turn a circle of fix feet diameter, I would prepare a chock, or platform, for the end of it, of that diameter, or a little more, compofed of clean-grained mahogany plank, all cut out of the fame log; which, when finifhed, to be about $1\frac{1}{2}$ inch thick, and formed in fectors of circles, fuppofe 16 to make the circle; the middle line of each fector lying in the direction of the grain of the wood, this will confequently every where point outward: the method of framing this kind of work is well known.

The way of getting a flip of brafs to anfwer the circumference of this platform is fuggefted in Mr. BIRD's account of conftructing mural quadrants. Let a parallelogram of brafs of about three feet long, and of a competent fubftance (fuppofe half an inch) to make it when finifhed about one-twentieth of an inch in thicknefs, be caft of the fineft brafs; and this to be rolled down till it becomes of fufficient length for the hoop, and about one-fifth part more. I would then cut off, from the whole length, fomewhat better than one-fixth part, the whole being fufficiently reduced to a thicknefs by the rollers. Perhaps no way will be more ready and convenient to get fuch a long ftrip of brafs reduced to an equal breadth, than the method prefcribed by HINDLEY; viz. by turning it upon the chock prepared; but I would not make it wider on one fide than the other, like the hoop of a cafk, as he defcribes, but exactly to fit the chock, when truly cylindric; for the internal elafticity of the brafs, in fo great a length, will be very fufficient for fitting it on tight enough, without any tapering. This I will now fuppofe done; and a pair of fteel *chaps*, as defcribed by HINDLEY, to be alfo prepared, and ready for grinding; which, by fuch a careful admeafurement as can eafily be made, will give the length of the hoop fufficiently near, on its firft preparation.

Method of forming a Pair of Straps as a Check to the Divisions.

The part firft cut off muft be again cut into two equal parts in length; which, for diftinction fake, I will call the *straps;* and which are to ferve as checks to every 60th and every 120th divifion of the circle.

A fteel plate, of about half an inch in breadth, the fame thicknefs as the ftraps, and in length equal to the breadth of the hoop plate, muft be foldered with filver folder to one end of each of the ftraps, by which means their length will be increafed half an inch by the fteel. An hole muft then be made through each fteel plate, of the fame fize as thofe through the chaps, and anfwerable to the middle of the ftraps; but fo near the border of the fteel, that when the chaps are put on, and adapted to the fteel hole, the next will fall through the brafs. The fteel plates muft then be hardened; and a pin being put through the two holes and the two plates, thefe muft be wrought to a right line in contiguity to each other; by this means the ftraight edge of each of the ftraps will be reduced to the fame diftance from the fteel hole: the hard fteel edges may be rectified by the grindftone, if neceffary.

This being done, not only the holes in the chaps, but the holes in the two fteel plates, applied to each other, like the two fides of the chaps, muft be refpectively ground together; not with a taper pin, as prefcribed by HINDLEY; but fo as not only to be cylindrical, but that the fame cylindrical pin fhall equally fit them all, and leave them fmooth and polifhed; which is a procefs no ways difficult to a curious artift, and of which therefore a minute defcription is unneceffary.

The chaps being then put upon one of the ftraps, with its ftraight edge uppermoft, and a pin put through the holes on the left-hand, and through the fteel hole in the ftrap under operation, the chaps muft be fet upright, fo that the line joining the centres of the holes fhall be parallel to the upper edge of the ftrap; the brafs plate mentioned by HINDLEY, between the chaps, as a guide for directing them always to that upright pofition, may be then adjufted and fixed to the infide of the chap next the operator *.

The performance of the enfuing part of this work fhould be at a feafon when the temper of the air is not very variable; rather above the mean temper (fuppofe 60°) than below it; but above all things the artift fhould be himfelf cool; that is, not in a ftate of fenfible perfpiration; and there fhould be a free circulation of air in the room. Things being thus conditioned in refpect to temperature, he may begin to drill the holes in one of the ftraps; the pin being firft put through the chaps and through the fteel hole of the ftrap; and the next hole, being drilled through the brafs with a common drill, that and every hole as it goes is to be finifhed with a taper broach, as prefcribed by HINDLEY; and he may then prove or finifh every hole by the ap-

---

* It would be well, previous to the drilling of the steel chaps, that another hole was drilled in the chaps, that should be somewhat above the upper edge of the straps, and in the middle betwixt side and side, to receive a *steady pin* therein, antecedent to drilling the main holes; for then a tempered steel pin, a little taper, will, by driving it in as far as necessary, constantly answer this purpose from first to last, so as to regulate the holes in grinding, to be truly opposite: proper holes should also be drilled for fixing the brass guide plate to one of the chaps.

plication

plication of a thorough broach, made fo full as to require a degree of preffure to force it through ; and this broach being a little tempered, and the holes quite hard, there will be no fear of injuring the fteel holes *.

Calling the hole in the fteel plates o, and obferving the time of beginning, you may proceed to drill 60 holes as prefcribed by HINDLEY ; and noting how long you have been about it, you may lay the work afide the length of time, equal to the time you took in drilling ; that any addition of warmth it may have acquired in handling or working may be again loft in a great degree †. After this paufe you may begin again, and go on to finifh 60 holes more ; that is, to the length of 120 holes from the beginning ; you then proceed in the fame manner with the other ftrap.

Method of drilling *the Hoop.*

You are now prepared to commence the work upon the long or *hoop-plate ;* and you proceed therewith, in forming the firft hole with the chaps, as before directed by HINDLEY, and this firft hole you call o. You then place the ftraps one on each fide the hoop, with their gaged edges upward, and put the pin through the holes denominated 60 upon the ftraps, and through the firft hole already made, and denominated o upon the hoop ; then, bringing the gaged edges of the fteel plates to be even with the upper or working fide of the hoop, you pinch them together in the vife, and drill and broach the hole through the fteel plates, which will make the hole, number 60, upon the hoop. This done, you put the pin through the left-hand hole of the chaps, and the hole marked o upon the hoop-plate firft made, and proceed to drill with the chaps to 59 holes inclufive, which will fill up the whole fpace from o to the 60th divifion before obtained.

---

* The steel holes in the chaps need not to be above one-twentieth of an inch in diameter ; and though it may be proper to make the steel plate, of which they are formed, one-tenth of an inch thick, in order to give the spring formed between them a convenient degree of stiffness, yet they may be reduced (by chamfering the outsides) to half that thickness.

† As there is not much occasion for the artist to touch his work, the effects of that may also be very much avoided by wearing thick gloves ; and the friction being but slight, and the work almost continually in the vise, the variation of temperature in the metals concerned cannot be sensible or confiderable.

You now again have recourfe to the ftraps, and placing them one on each fide the hoop-plate, you put the pin through the 120th hole of the ftraps, and through the hole marked o upon the hoop-plate; and regulating the fteel plates to the hoop-plate as before, you drill and form a hole with the fteel plates, which will correfpond with the 120th hole upon the hoop-plate; and afterwards filling up the 59 holes wanting, by means of the chaps, you then have all completed to the 120th divifion, which is one-twelfth of the whole circle.

You then proceed, in like manner, with another fet of 120 holes; that is, placing the 60th hole of the ftraps to the 120th hole of the hoop-plate, and from it producing the 180th hole; you, in like manner as before, fill up this 60 with the chaps; and afterwards placing the 120th hole on the ftraps in the 120th hole on the hoop-plate, you will obtain the 240th hole; fo that filling up this laft fet of 60 divifions, you have obtained 241 holes, including 240 fpaces or divifions of the hoop; and repeating this procefs ten times more, you will, in like manner, obtain 1441 holes, comprehending 1440 fpaces *. And this procefs being carried on in temperate weather, the manner of working produces twelve fimilar operations, wherein the materials and tools concerned will not only be fubjeďt to very little change of temperature, but that change, whatever it is, will be nearly fimilar in each fet of 120 holes: we may therefore infer, that the greateft inequality, or indeed any that can be fenfible, muft be at every 60 divifions, that is, between the 59th and 60th, and between the 119th and 120th, both which will be equally repeated 12 times, in the whole length which is to compofe the *circumference of a circle,* and which will thus be checked thereby 12 times in the circumference, and 12 times more at the intermediate diftances; that is, with 12 mafter checks, and 12 fubordinate ones, in the whole round.

It is proper here to obferve, that in M. Roemer's method even fixty divifions could fcarcely be trufted in an affair of great accuracy, on account of the objeďtions already made, arifing from the points having fuch flight hold in the furface of the brafs; but here the parts are held fo exceedingly firm, and the operation carried on with fo much power, that any fmall inequality in the hardnefs of the brafs, or irregularity of furface,

---

* It will be proper, for reasons hereafter to be mentioned, to continue the divisions to 20 holes more, making in the whole 1461 holes.

cannot

cannot be fuppofed to affect the place of the centre of the hole ; nor will any fmall inequality that may be fufpected from the wear of the fteel holes fenfibly affect the *centre* of the hole, to which every thing is ultimately referred.

Method of joining *the Hoop.*

A more happy thought than that of HINDLEY's, for joining the two ends of the hoop, could fcarcely have been wifhed for, in regard to preferving the fame equality of the fpace between the holes contiguous to the joint, as in the other parts : for though, geometrically fpeaking, the two *faddle* plates, in which the little cylindrical bolts are fixed, for bringing the terminating holes of the hoop plate to their due diftance, being one applied within the hoop, and the other without, will belong to circles of different *radii* ; yet this difference being exceedingly fmall in fuch thin metal, and fo great a radius, and one being as much too big for the hoop as the other is too little, when the bolts are put in, and the hoop in that part fet nearly to a circle by a mould ; the mean between them affumed by the hoop, from the elaftic compreffibility of the materials, will be the truth.

It muft, however, be remarked, that in the ufe of the ftraps, the joining of the hoop fhould not be made at any part betwixt an 119th and an 120th divifion, as fome inequality muft be fuppofed there, unlefs the faddle plates were adapted thereto. The method the moft eafily practifed, will be to continue the divifion upon the hoop, about twenty more than the completion of the number intended to form the circle, and to cut off all the overplus ones at the beginning.

The faddle plates I would recommend to contain ten holes each ; fo that if the divifions are carried on to twenty more than what will be contained in the circle, there will be a piece containing twenty to cut off ; and this again being cut in the middle will afford ten holes to make each faddle plate ; fo that there will be a place for a bolt on each fide the joint, and then putting a bolt through every other hole, there will be three bolts at an end.

The pieces deftined for the faddle plates, thus obtained, being broader than can be admitted when put to this ufe, I would advife to divide the breadth of the plate into three equal parts ; and with a cutting hook (which perhaps will be attended with the leaft violence in the feparation) to feparate the two outfide pieces from the middle piece :

by

by this means the two faddle plates (though double) will occupy one-third only of the breadth of the hoop in the middle; and two of the pieces cut off being applied, one on each fide of the faddle plate on the outfide, will anfwer in like manner for the *rivet* plates.

The laft operation to complete the joining of the hoop is the putting on the rivet plates: to complete this, I would advife a piece of brafs, of three or four inches in length, to be filed fo as to anfwer to the infide of the hoop, when reduced to a true circular form; and being three-eighths, or one-half an inch in thicknefs, to file the oppofite fide fomewhat nearly concentric thereto; apply the middle of its convex arch to the infide of the hoop at the joint, and then bring on the middle of one of the rivet plates to the joint of the hoop, confine the three together by a couple of narrow-chapped hand vifes, leaving a fpace between them capable of receiving a couple of pins as rivets on each fide the joint; the holes for the rivets are then to be drilled through all, and a little fmoothed with a broach at their entry, into which fmooth taper pins are to be driven; not with violence, but moderately, that no fenfible ftretching of the folid parts may take place thereby; then cutting off and fmoothing the heads, fhift the vifes fo as to receive another couple of holes, and a third couple in the fame end of the hoop; and proceed progreffively in the fame manner, from the middle to the other end of the rivet plate; then gently feparate the internal brafs mould with a thin knife, or fuch like inftrument; and cutting off, and very lightly rivetting the inner ends, proceed to fix the other rivet plate, in the fame manner, on the other fide: by this means the hoop will be firmly joined in the very pofition given it by the faddle plates and mould. Thefe plates may then be removed, the infide of the hoop cleared and fmoothed, if neceffary; and the outfide will have the middle part clear where the divifions lie, and that without fenfible lofs or gain in the juncture.

Method of transferring *the Divisions of the Hoop to a dividing Plate.*

The hoop being thus refitted for the chock, that fhould be turned down to leave a fhoulder on one fide, that the hoop, now reduced to an equal breadth, may be forced againft it; and the divifions, being equally diftant from one of its edges, will be all found in a circle, as if turned upon it. It fhould be very carefully fitted to the chock, that it may go on with a fufficient degree of tightnefs, and without the neceffity of much forcing; and it will be no inconvenience now, if it goes on upon a very flight

degree

degree of taper of the chock, as the internal fpring of the materials will eafily accommodate it to this fhape without any injury to its general truth: a flight degree of a
groove fhould be turned in the place where the divifions will come, that any conical
pin, that is to ferve as an index, let drop into the divifions or holes, may not, by
reaching through this thin plate, abut upon the wood, rather than upon the fides of
the holes; and thus this hoop is made into a wheel of 1440 equal divifions, moveable round upon its own axis, whereon it was formed.

Againft the time that this is completed, there muft be prepared a flat circular plate
or wheel of brafs, the rim of which fhould be of about $3\frac{1}{2}$ inches breadth, and about
two-tenths of an inch in thicknefs when finifhed, to make a dividing plate; the external diameter of this is to be fuch, that when laid flat upon the furface of the
mahogany platform, its extreme edge will exceed the diameter of the hoop by about
half an inch all round.   There muft alfo be prepared brafs arms (fuppofe eight in
number) of an equal fubftance with the outer rim, and all connected with a circular
plate in the middle; and, the whole of this work being framed beforehand, is to be
let on flat upon the mahogany platform; whofe face is fuppofed to be turned truly
flat, and fufficiently affixed with fcrews: in this fituation, the outward edge is to be
turned, and the outward face of the rim turned flat.   The *centre plate*, which may
be about twelve inches diameter, is alfo to be turned as flat as poffible, and a centre
hole, of about half an inch diameter, to be very carefully turned therein.

A piece of clean, ftraight-grained, well-feafoned mahogany, of about two feet long,
three inches thick, and five or fix inches broad, is then to be well affixed to fome
part of the general frame of the lathe, which muft now have its pofition altered, fo
that the platform will become horizontal; and therefore the frame fhould be originally
made with this view *.   The piece of mahogany is to be affixed fo that one of its
larger faces fhall be in a parallel plane to the face of the platform, and fo low as to
clear the under fide of the platform in its rotation; and fo far diftant from the centre,
that an index may be fixed upon this upper face of the piece of wood, fo as conveniently to drop into the holes of the hoop; while the common *cutter frame* of a
clock-maker's engine fhall be firmly attached upon the fame face of the wood, and fo

---

* After changing the position of the lathe, the collar of its mandrel should be removed, and the neck
made to move within three planes, so as to preserve an exact centre, in the manner of an *equal altitude*
instrument.

fixed

fixed as to cut the edge of the dividing plate into teeth, anſwerable to the ſeveral diviſions of the hoop. The teeth need only to be cut with a common cutter, making a parallel notch : and here it will be proper to obſerve, that not only both the index and cutter are to be founded on the ſame piece or baſe of wood; but that the nearer they are together, the more free they will be from the effects of all variations of expanſions by variations of temperature *.

The equaliſing the Teeth of the dividing Plate by grinding.

The object of transferring the diviſions of the hoop to the teeth of the dividing plate, is ſtill farther to equalize the teeth by grinding; eſpecially thoſe that, falling within the compaſs of each ſet of 120 diviſions, may be ſuppoſed, if any, to be mended thereby; but as it may be incommodious to conſtruct a curved ſcrew, of ſuch a length and ſize, in HINDLEY's method, as would be ſufficient for the purpoſe; I would propoſe to uſe two ſcrews of braſs, cut from a cylinder in the way ſet forth by Mr. RAMSDEN, each of which, with a very little grinding upon this large circumference, would lay hold of ten or twelve teeth together. I would place the two ſcrews, that is, their middles, to be ninety diviſions aſunder; of conſequence, when one of the ſcrews is between the 59th and the 60th, or between the 119th and 120th diviſion of each ſet, the other will be in the middle of the ſpace divided by the chaps only †.

The threads of theſe ſcrews I would adviſe to be cut a little taper, ſo that as they grind in, they may fill the notches of the teeth; which alſo, by this means will acquire a little tapering towards their extremities; and by cutting the notches parallel,

---

* It is proper to obſerve, that as it may be impracticable to get the rim of the dividing plate caſt of the proper size, in one entire piece, it will be very practicable, if caſt of a leſs size (ſuppoſe half), but of a ſufficient thickneſs, to roll it down; and by having the outward edge originally thicker than the inner, in the proportion of the *radii*, it may be ſo managed by the rollers as to be of an equal thickneſs when brought to its proper size. But the arms and centre plate ſhould be of the ſame metal, rolled in the ſame degree.

† The beſt way of giving an equal motion to thoſe two ſcrews, ſeems to be by a detached axis carrying two common flat wheels; one acting upon a like flat wheel, upon the axis of one ſcrew, and the other, in the ſame manner, upon the other; and applying the pulley for communicating the power to the middle of the detached axis between the two wheels, the ſpring or twiſt will be equal both ways; ſo that in turning the contrary way round, they will ſtill be in equal advance.

as

as I have mentioned, the true ground part will always be certain of being at the extremity.

When the fcrews have been ufed in grinding till they are found to have the effect of a perfectly equal and eafy rotation all round, and all the teeth reduced to a fenfible taper, and regular bearing, I would then totally remove the fcrews from the fquare block of wood, upon whofe upper face I fuppofe them to have been mounted; in like manner as I fuppofe the index and cutting frame to have been removed, to make room for the mounting of the fcrews. I now confider the teeth of the dividing plate, fo formed, as having all the equality that the prefent known ftate of human art has pointed out; and the whole convertible upon the axis or mandrel upon which it has been originally formed, and the central hole of the plate concentric therewith: I therefore confider the ground faces of the teeth of the plate as the actual divifions. It now remains to fhew how they are to be transferred, to form the divifions of an inftrument.

Preparation *of the dividing Plate for* graduating *Instruments.*

If a fmall cylinder of hard fteel is duly polifhed, and made of a fize fo as juft to chock in betwixt the extremities of the teeth, then the centre of that cylinder will be a fixed point, in refpect to the circumference of the wheel: if another cylinder is applied in like manner, at the diftance of a number of divifions (fuppofe it a prime number, fo as to crofs all former divifions, viz. 17 or 19), then the middle of the line joining the centres of the two cylinders will remain in the direction of the *same radius,* though one of them fhould force in a minute quantity further than the other; and if a point is affumed in the direction of a tangent to a circle at this middle point, then though both the cylinders fhould drop in a minute quantity further at one time than another, yet the middle point would remain at the fame diftance from the point in the tangent; provided that point was removed to a competent diftance, that is, to five or fix inches. On this principle I would conftruct an index, the two cylinders being fixed in a frame, convertible about the middle point, and to be centered in the end of the lever, reprefenting the tangent; then this lever being again convertible about the point in the tangent line, the middle point would always have a fixed diftance from the point in the tangent, and there hold it fteadily faft; the tangent point being placed upon the fixed block beforementioned.

Ufe

Ufe *of the dividing Plate in the* Graduation *of Instruments.*

Our dividing plate is now ready for the reception of an inftrument; fuppofe it a quadrant, whofe radius, however, muft not exceed the radius of the dividing plate : It is to be laid upon the face of the dividing plate, and a weight, or weights, equivalent to that of the quadrant, is placed on the oppofite fide, to balance it. It muft alfo be fuppofed, that the quadrant is made with a view to be divided by this engine ; and confequently, that the central cylinder is fo well adapted, and nicely fitted to the centre hole of the quadrant, that the centre cylinder can be removed, in order for the limb to be divided, and again replaced, without fenfibly altering its centre. This being the cafe, let a piece of metal be turned, to apply to the quadrant, perfectly like its centre cylinder at the upper end, and turned nicely to fit the central hole in the dividing plate, at the lower end ; then, the quadrant being fixed with proper faftening fcrews, I would cut the divifions with a beam compafs ; and, if a fixed point is affumed, viz. the centre of the tangent point for the index ; then the beam compafs being always opened to the computed length of the tangent of the circle of divifions, it will be fufficiently near for cutting the divifions, fquare to the circular arches between which they are placed.

It will alfo be proper (to prevent unequal expanfions) that the beam of the compafs fhould be formed of a piece of clean-grained *white fir ;* and that the length between the points be inclofed in a tube of tin or brafs ; without touching the beam, except at the terminations, which will in a great meafure protect it from both alteration of moifture, and of heat from the body of the artift, during the operation.

It will be likewife proper to have a lever, or fome equivalent contrivance, to bring the dividing plate forward ; that after lifting the little cylinders out of the divifions, and refting them upon the tops of the teeth, they may be brought gently forward with an equal drag, and ultimately fnap in between the teeth, by the ftrength of the fpring commanding the index ; by this means the drag of the friction of the whole will be conftantly the fame way.

Conclusion.

Now, if, as it has been fhewn, a quadrant of any radius may be read off to the 4000dth part of an inch, then this quantity upon a radius of three feet will not be fo

much

much as $1\frac{1}{2}$ fecond; and as the whole of the procefs is carried on by contact, in which a greater error than that of a 60,000dth part of an inch cannot be admitted in any fingle operation, I fhould affuredly expect a three-feet quadrant, fo divided, to be true in its divifions, and read off to at moft two feconds.

But, after all, in an inftrument like this, I fhould expect the greateft fource of error to be in the want of perfect coincidence of the centre of the divifions with the actual centre upon which the index revolves; and therefore, that if, inftead of a quadrant of three-feet radius, a complete circle of five feet diameter was divided, and its divifions read off from the two oppofite points (taking the mean), then the errors of the centre will be wholly avoided. For this reafon, I am very clearly of opinion, that the fagacious propofition of Mr. RAMSDEN, to ufe circles inftead of quadrants, or other portions of circles, will bid much the faireft for perfection in actual practice; and that his ingenious method of making them both ftiff and light, by the ufe of hollow conical tubes by way of fpokes, in the manner of a common wheel, will enable him to mount them of five feet diameter, upon hollow axes, in the nature of a *transit*. By this means we fhall have all the good properties of both the quadrant and tranfit united in one inftrument; and obfervations both of right afcenfion and declination, through the very fame telefcope, as long fince attempted by M. ROEMER; and to a degree of perfection and certainty, in point of declination, hitherto unattainable by the largeft inftruments that have yet been made.

N. B. In matters of very nice determination, fmall circumftances often come to be of confequence; and it is in this view that I mention what follows. It was a practice of HINDLEY's of many years ftanding, and fince followed by myfelf and others, wherever he made any ufe of the *vernier*, to lay the vernier plate in the fame plane, or cylindrical furface continued, whereon the principal divifions are cut. It is of equal utility, though the vernier be rejected, to lay the index ftroke in the plane of the divifions. In this way the divifions being by convenience upon the external border of the limb *, *two* fets of divifions are thereby rendered incommodious; but thofe that

---

* It has been objected, that laying the divifions upon the extreme edge of the limb of the inftrument fubjects it to injury: but, to obviate this, in an HADLEY's quadrant made for me, by my direction, by the late Mr. MORGAN, in the yesr 1756, wherein the vernier is laid even with the divifions, those are protected by a projection of the folid part of the limb, beyond the divifions; a *Rabbet* being funk in the edge of the limb, to clear the vernier.

wifh

wifh two fets, as a check, will in a great meafure aid themfelves, by reading from two different parts of the fame fet of divifions ; which is very eafily provided for, by putting an additional ftroke upon the index plate, at the diftance of 9, 11, or any prime number of divifions to 19, 23, or more ; and reading off from that ftroke alfo ; as before recommended for great quadrants, where the vernier is propofed to be rejected * : fo that they will thereby be mutually checked by divifions that had no correfpondence in their original formation.

---

* I would not have it thought, from my proposal of rejecting the vernier, that I have any quarrel with it ; I think it a very simple and ingenious contrivance, where it is properly applicable ; that is, where the strokes of the vernier, or their estimated halves, are sufficient for all the precision required or expected from the instrument, as in HADLEY's quadrants, theodolites, &c. but where still more minute divisions are required than can easily be had by estimation from the vernier ; to do this by a screw, as a *supplement* to the vernier, appears to me in the light of bringing a more accurate tool to supply the deficiencies of one less accurate; when the former might, with more propriety, supply the place of the latter altogether.

REMARKS

REMARKS on the different Temperature of the Air at Edystone, from that observed at Plymouth, between the 7th and 14th of July, 1757, by Mr. JOHN SMEATON, F. R. S.

Read January 12th, 1758.

SIR,

ON reading of Dr. HUXAM's letter at the laft meeting, fome obfervations occurred to me, concerning the different temperature of the air, which I had obferved at the Edyftone, from what had been obferved by the Doctor at Plymouth, between the 7th and 14th of July laft ; which, having been defired by fome members to be put into writing, I beg leave to trouble you with the following :

Edyftone is diftant from Plymouth about fixteen miles, and without the head-lands of the found about eleven.

The 7th and 8th were not remarkable at Edyftone for cold or heat ; the weather was very moderate with a light breeze at eaft, which allowed us to work upon the rock both days when the tide ferved.

About midnight, between the 8th and 9th, the wind being then frefh at eaft, it was remarkably cold for the feafon, as I had more particular occafion to obferve, on account of a fhip that was caft away upon the rocks.

The wind continued cold the 9th all day, which was complained of by fome of the fhipwrecked feamen, who had not time to fave their clothes; and fo frefh at eaft as prevented our going near the rocks, or the wreck; and fo continued till Sunday the 10th ; when feeing no profpect of a fudden alteration of weather, I returned to Plymouth in a failing boat, wrapped up in my thick coat. As foon as we got within the head lands, I could perceive the wind blow confiderably warmer, but not fo warm as to make my great coat uneafy. Having had a quick paffage, in this manner I went home, to the great aftonifhment of the family to fee me fo wrapped up, when they were complaining of the exceffive heat: and indeed it was not long before I had reafon to join in their opinion.

This

This heat I experienced till Tuefday the 12th, when I again went off to fea, where I found the air very temperate, rather cool than warm, and fo continued till Thurfday, the 14th.

In my journal for Wednefday the 13th, I find the following remarks, viz. "This evening's tide," (from 6 A. till 12 A.) "the wind at eaft, but moderate, with frequent flafhes of lightning to the fouthward. Soon after we got on board the ftore veffel, a fquall of wind arofe from the fouth-weft on a fudden, and continued about a minute, part of which time it blew fo hard, we expected the mafts to go by the board; after which it was perfectly calm, and prefently after a breeze returned from the eaft."

And in the journal of the 14th is entered, " This morning's tide," (viz. from 1 M. to 1 A.) " the air and fea quite calm."

Hence it appears how different the temperature of the air may be in a fmall diftance; and to what fmall fpaces fqualls of wind are fometimes confined.

It may not be amifs further to obferve upon this head, that once, in returning from Edyftone, having got within about two miles of the Ramhead, we were becalmed; and here we rolled about for at leaft four hours; and yet at the fame time faw veffels not above a league from us, going out of Plymouth Sound, with a frefh of wind, whofe direction was towards us, as we could obferve by the trim of their fails; and as we ourfelves experienced, after we got into it by tacking and rowing.

I am,

SIR,

Your moft humble Servant,

J. SMEATON.

Furnival's Inn Court,
12th Jan. 1758.

AN

AN ACCOUNT of the Effects of Lightning upon the Steeple and Church of Lestwithiel, Cornwall; in a Letter to the Right Honourable the Earl of MACCLESFIELD, President of the Royal Society, by Mr. JOHN SMEATON F. R. S.

Read April 21ft, 1757.

JANUARY 25th, 1757, about five o'clock in the evening, returning home from the Edyftone works, near Plymouth, I obferved four flafhes of lightning, within the fpace of fix or feven minutes, towards the weft, but heard no noife of thunder, being diftant thrty miles. A fews days after I was informed that the fame evening the lightning had fhattered the church of Leftwithiel in a very furprifing manner.

The 1ft of March I was at Leftwithiel; they had then begun to repair the damages; but had not made fuch a progrefs but that the principal effects were equally obfervable as at firft I obferved, and was informed as follows :

At the time before mentioned, the inhabitants were alarmed by a violent flafh of lightning, accompanied with thunder fo fudden, loud, and dreadful, that every one thought the houfe he was in was falling upon him; almoft every one being within doors, on account of a violent fhower of rain, which preceded the lightning; fo that nobody faw or heard any thing of the mifchief done to the church, till it was obferved accidentally after the fhower.

The fteeple is carried up, plain and fquare, to about forty-nine feet, with a kind of flate ftone, rough cafted on the outfide; upon which is formed a very elegant octagon gothic lanthorn, about nine feet high, and thereon a ftone fpire about fifty-two feet height, with a fpindle and vane rifing about three feet above the ftone; fo that the whole together was about one hundred and thirteen feet. Each face of the lanthorn finifhes above with a fort of a gothic pediment, with a little pinacle upon each, feparated from the body of the fpire.

I will not affirm, that the lightning entered in at the fpindle or vane at the top, but will fuppofe it, for the fake of methodizing the facts. The vane was of copper-plate, which, being turned round, and rivetted, made a focket to turn upon. The fpindle did not reach through the focket, but the weight of the vane refted upon the top of the fpindle, the top of the focket being clofed. About the vane were many acute angles, and fome almoft fharp; but I did not obferve any pointing directly upward. The vane

was

was much bruifed, which might be occafioned by the fall ; but the focket was rent open as if it had been burft by gun-powder, and in fuch a manner as I cannot conceive could be occafioned by the fall.   Under the fpindle, that carried the vane, was a bar about four feet long and one inch fquare, that paffed through the centre of feveral of the uppermoft ftones fucceffively, in order to unite them more firmly together, and was run in with lead ; all which furrounding ftones were broke off, except one, which, together with the bar, fell down within the tower.

The fhell of the fpire, as far down as thirty-five feet from the top, was no more than feven inches thick, and the courfes about the fame heighth ; fo that fcarce any one ftone in the fpire could weigh more than thirty or forty pounds ; but they were joined together at the ends with mortoife and tenon, in a curious manner.   Above twenty feet of the upper part was entirely thrown down, and difperfed in all directions ; and, as I was informed, fome pieces were found at the diftance of two hundred yards.   A great many ftones fell upon the roof of the church ; and feveral made their way through both roof and ceiling down into the church, breaking the pews and whatfoever they fell upon. Six feet ftill lower the fpire was feparated, the wefternmoft half being thrown down ; the eaftern half was left ftanding, but disjointed, and in fo critical a pofture, that it feemed ready to fall every moment, fo that this was ordered to be taken down immediately ; and likewife to fix feet below, the work being found remarkably fhattered. In this condition it was when I faw it.   The whole of the fpire I found much cracked and damaged, but the remainder of the feven inch fhell fo greatly, that there feemed fcarcely a whole joint.

The pediments over every face of the lanthorn were damaged more or lefs, but the whole afhlering of that to the N. W. was torn off from the inner wall ; to which it was connected.   At firft fight this might feem to be done by the falling of the ftones from above ; but I was convinced to the contrary, by obferving, that feveral of the pediments were damaged, and even ftones ftruck out, where the little pinacles above them were left ftanding.

About the top of the lanthorn is a bell for the clock to ftrike on : it is hung upon a crofs-bar, with gudgeons at each end ; the whole being fufpended to a beam laid acrofs the tower.   The crofs bar was fo bent, that the clock hammer could not touch the bell by above two inches.   This could not be done by falling of ftones, becaufe the beam would defend the bell from receiving any ftroke in the direction to which the crofs bar was bent.   As to the wire that drew the hammer, as I was informed, not one bit of it could be found.

The

The bells (four in number) for ringing, hung in the fquare part of the tower, below the lanthorn, two above and two below : the wheels of every one were broke to pieces, and one of the iron ftraps, by which they are faftened to the yoke, unhooked, and as appeared to me could not be replaced without great force or unloofing. Whether thefe accidents were occafioned by the lightning or the falling of ftones I leave undetermined. In the floor under the bells was placed the clock, cafed up with flight boards. The verge, that carries the pallets, was bent downwards, as if a ten pound weight had fallen ten feet right upon it. The crutch, that lays hold of the pendulum, looked as if it had been cut off by a blunt tool, and heated by the blow, till it was coloured blue, at the place where it was cut. It turned at a right angle, and might be about $\frac{4}{10}$ of an inch broad by $\frac{2}{10}$ thick. As to the pendulum, which hung pretty near the wall, the upper part of the rod was ftruck with fuch violence againft the wall, that a fmart impreffion thereof was made in the plaifter : and near the upper part of the impreffion appeared a fhady ring, of a blackifh colour, fomething like as if a piftol had been difcharged of powder, and the muzzle held near the wall. The cafing of the boards round the clock remained unhurt.

In this ftory, on the north and fouth fide, are two narrow windows or air loops; againft the upper part of which, on the outfide, were fixed the timber dials belonging to the clock, both of which were blown off and broke to pieces, poffibly by the fall : and not only that, but part of the ftone jambs were broke out alfo, near to where the rod paffed, that carried the hands. In this ftory alfo was a fort of window or air loop on the eaft fide, that had communicated with the church, but was ftopped up with lath and plaifter; alfo feveral putlock holes for the fcaffolding, which had gone through the wall into the church, but were ftopped up with ftone, and plaiftered over; all thefe were forced out into the church and the plaifter torn from the wall.

In the ground ftory of the tower or bellfrey, the fouth and north entrances were fhut with wooden doors. The upper part of the eaftern entrance, that communicated with the church, was made up with lath and plaifter; and before it, in the church, are feats raifed one higher than another; fo that the floor of the feats next the wall was half up the door way; confequently the vacuity under the feats lay open to the bellfrey.

About the middle of the wefternmoft fide one of the paving ftones, about one foot fquare, and one and a half inch thick, was thrown up, and a hole pierced into the wall, rather below the level of the pavement, into which one might put three fingers, on the oppofite fide, the fouth weft angle of the middle buttrefs, which together is eight feet; but the hole was too rugged and crooked to put any thing

through

through. Befides this hole the wall was pierced in feveral places, and the plaifter thrown off both within and without. One place within about four feet above the floor, was a hole of about fourteen inches fquare, pierced fix inches in the wall; and fo nearly fquare, that I enquired whether it had not been made by art, but was affured of the contrary.

The north and fouth doors of the tower were both blown out, and broke in m an pieces. Many of the arch ftones over both doors were disjointed and difplaced: two of the ftones making the jamb of the fouth door were forced quite out, and one of them broken.

The vaulting of the eaft door way was plaiftered underneath: the plaifter was fprung from the ftone in 30 or 40 places, like as if a bar of iron had been drove from above through the joints of the ftone, and thereby forced off the plaifter, with its end. The lath and plaifter partition, which ftopped up the upper part of this door-way, was forced into the church, and the wainfcotting making the back of the laft feat was torn from the wall from end to end. Some part of the vapour feems to have made its way through the cavity under the feats; for moft of the boards, compofing the rife of the fteps from feat to feat, were blown out forwards; and feveral pannels of wainfcot at each end of the feats above mentioned, were forced out and broken. Hence, the vapour feems to have divided itfelf into three branches, one moving directly forward to the eaft window, being 13 feet wide, and about 20 feet high, confifting of five principal lights divided by ftone mullions: two of the lights were in a manner wholly deftroyed, and feveral large holes in thofe remaining; the glafs and lead being carried outward, like as if an harlequin had leaped through the window. The north window, fronting the broken pannels of the feats, was very much fhattered: but the fouth window oppofite had fcarce a whole pane left.

It is farther to be noted, that almoft all the lights in the church, though not broken, were bagged outward.

N. B. It was faid in the London Papers, that the organ was entirely fpoiled: it is certain there is not, nor ever was, any organ in this church.

FINIS.

Printed in the United States
By Bookmasters